THE CAMBRIDGE LIBRARY OF
MODERN SCIENCE

PARASITIC ANIMALS

PARASITIC ANIMALS

BY

GEOFFREY LAPAGE

M.D., M.Sc., M.A.

*Lecturer in Animal Pathology in the University of Cambridge;
formerly Head of the Zoology Department, University
College, Exeter, and late Lecturer in Zoology,
University of Manchester*

CAMBRIDGE

AT THE UNIVERSITY PRESS

1951

CAMBRIDGE
UNIVERSITY PRESS

University Printing House, Cambridge CB2 8BS, United Kingdom

Cambridge University Press is part of the University of Cambridge.

It furthers the University's mission by disseminating knowledge in the pursuit of education, learning and research at the highest international levels of excellence.

www.cambridge.org
Information on this title: www.cambridge.org/9781107496613

First published 1951
First paperback edition 2015

A catalogue record for this publication is available from the British Library

ISBN 978-1-107-49661-3 Paperback

CONTENTS

PREFACE

The title of this book, *Parasitic Animals*, has been chosen for
two main reasons. First, it tells the reader that the book is
about living organisms which have adopted the parasitic
way of living and are therefore called parasites. Secondly,
it tells him that the book deals only with animals that are
parasitic. It does not, that is to say, consider, except
incidentally, the other kinds of living organisms which may
be parasitic, namely, the viruses and the plants. Certain
parasitic viruses cause serious diseases of man and of other
animals and plants. The parasitic plants include, not only
such highly organised species as the mistletoe, but also all
the organisms generally known as bacteria, whose injurious
effects need no explanation here.

A single volume could not give an adequate account
of all these three different kinds of parasitic organisms.
Even if we consider, as this book does, the parasitic animals
alone, it is not possible to give in a book of this size more than
an introduction to our knowledge of these organisms and an
outline of the important problems which the study of them
raises. Some of these problems, fascinating as they are to
biologists, will not interest other people. The biologist,
always interested in the animal as a living unit in relationship to its environment, finds the parasitic animal just as
interesting from this point of view as animals which live in
other ways. He studies with delight their intimate and
often complex relationships to their various environments
and finds these just as beautiful, remarkable and effective
as those discovered in other fields of biological work. Yet,

shadowing his delight, and often destroying it completely, is the knowledge he gains that parasitic animals cause so much human and animal suffering, and kill so many men and women and animals and plants. Out of this shadow arises the desire to fight and control the parasitic animal and to mitigate or to remove the death, suffering and economic losses that it inflicts. Some of the methods which he uses for this purpose and the biological principles upon which these methods are based are indicated in the course of this book.

The book, however, is not in any sense a text-book of economic parasitology. Its aim, on the contrary, is to give, from the biological rather than the economic point of view, an outline of the general principles which govern the lives of parasitic animals. These principles are illustrated largely by reference to species of parasitic animals which attack either man or his domesticated animals, so that the reader will gain, either by implication or direct reference, some idea of the great extent and variety of the losses inflicted by parasitic animals on human civilisation. But other species which do not harm man and his civilisation are also discussed, so that the biological basis of economic parasitology is not, the writer hopes, obscured. It is, indeed, one of his aims to show that the control of parasitic animals depends upon the biological study of them as animals living in a particular way.

G. LAPAGE

CAMBRIDGE
May 1950

ACKNOWLEDGMENTS

The author is glad to have this opportunity of acknowledging the debt to all the biologists whose disinterested work has made it possible to write this book. To Dr Charles Wilcocks, Director of the Bureau of Hygiene and Tropical Medicine, and Dr M. Abdussalam, of the Animal Husbandry Research Institute, Peshawar, he is grateful for constructive criticism of the manuscript and correction of some errors of fact. He is also grateful to all the members of the Cambridge University Press who have expended so much care on the production of the book. For the book as it now is, and for any errors that may still remain in it, the author alone is responsible. Thanks are also due to Mr D. W. Muncey, who read the manuscript from the point of view of a reader without special training in biology and made useful criticisms of it.

The line drawings reproduced in figs. 22–24, 33–39, 54 and 55 were drawn by Miss B. Pickering from the author's sketches. For original photographs the author is grateful to Dr David Robertson, Zoology Department, North of Scotland College of Agriculture, for figs. 1, 2, 12, 65, 67, 68 and 90; Dr H. Spencer, St Thomas's Hospital, London, for fig. 73; Dr A. H. Hamilton, St Thomas's Hospital, London, for figs. 69, 70, and 92; Mr R. K. Farmer, Boots Pure Drug Co., Ltd., Veterinary Science Division, for fig. 66; Mr G. H. Werts, Imperial Chemical (Pharmaceuticals) Ltd., for figs. 11, 30, 32, 85, 87–89, 91 and 93; Dr C. Horton-Smith, Animal Health Trust, Poultry Research Station, Houghton Grange, Huntingdon, for figs. 25, 26 and 28; the Cooper Technical Bureau, Berkhamsted, for figs. 13–16, 86, 94–99; and Mr W. J. Smith, Institute of Animal Pathology, University of Cambridge, for figs. 3, 4, 27 and 29. The sources of the other illustrations are gratefully acknowledged in the list of illustrations.

G. L.

CAMBRIDGE

May 1950

ILLUSTRATIONS

CHAPTER 1

WHAT IS A PARASITIC ANIMAL?

A parasitic animal is not a particular species of animal, but an animal which has adopted a certain way of living. A great many different kinds of animals have, in the course of evolution, become parasitic. The single-celled malarial parasite, which brings so much suffering to man in certain parts of the world, is one kind of parasite; the mosquito, which injects it into human blood, is another kind; the liver fluke which harries, and may kill, our sheep and cattle, the hookworms which suck the blood of man and animals, the tapeworms, the leeches, the fleas and ticks, the hagfishes which eat the flesh of other fishes in the sea, together with many other species of animals which do no harm at all to man and his civilisation—all these are also parasitic. The word 'parasitic' therefore refers to the way of life of many very different kinds of animals. What does it mean precisely? We must know its precise, its scientific meaning. If we do not, we shall quickly lose ourselves in a maze of error and misunderstanding.

Unlike many other biological terms, the word 'parasite', and its adjective 'parasitic', have been taken into the every-day language of men and women, and have, in the course of common usage, acquired emotional and moral connotations with which science—and therefore biology—has nothing whatever to do. In our dictionaries, and in ordinary language, the words 'parasite' and 'parasitic' refer to a way of living which we usually deprecate and despise—the life, that is to say, of the sycophant, the hanger-on, the person

who does no honest work, but lives by depending upon the
efforts of others, or who goes even further along what we
regard as a downward path and lives by actually preying
upon and harming his fellow human beings and their
society. This is not, however, the meaning which the biologist gives to the word. His outlook is scientific and,
because it is so, he does everything in his power to remove
from his studies all human likes and dislikes and all human
moral judgements. He neither despises nor admires, likes
nor dislikes, condemns nor approves, the parasitic organism.
He studies it and its way of living as dispassionately as he
can, seeing parasitism as one of the various ways of living
practised by different kinds of animals.

Adopting this scientific attitude to parasitism, the biologist recognises it as one of several kinds of *associations* which
animals form, either with other animals or with plants. The
ecologist, who studies the habits and lives of living organisms
in relation to other living organisms and to their environments, recognises that these *animal associations*, as he calls
them, can be classified in various ways. They can, for
instance, be divided into:

(*a*) Associations of individual animals all of which belong
to the same zoological species, such as herds of cattle, flocks
of geese and sheep, colonies formed by corals, and the
elaborately organised communities of some species of ants,
bees and wasps; and

(*b*) Associations of individual animals which belong to
different zoological species, with which may be included
associations which animals form with plants. To this second
category parasitism belongs.

Alternatively, however, animal associations can be use-
fully classified according to the reason for their formation.

We can postulate, for instance, that some forms of animal associations arose from the need experienced by all kinds of living organisms to obtain an adequate food supply. This need, we may suppose, impelled, in ancient times, different kinds of animals to share a common supply of food, and in this way there arose the association of different kinds of animals to which we nowadays give the name *commensalism*.

The word commensalism means literally 'eating at the same table'. It was originally given to associations of animals of different kinds which share each other's food. An example of this kind of association which will be familiar to the reader is the association of some species of hermit crabs with sea anemones which feed upon scraps of food not eaten by the crab. The term commensalism has been extended to include associations which confer, not only, and not even necessarily, the benefit of a common food supply, but other forms of benefit, such as shelter, warning of the approach of enemies, transport about the environment and other advantages which increase the commensal's chances of survival. Thus the bird called the ox-picker perches upon the backs of rhinoceroses, elephants and other African mammals, feeding upon the lice and ticks which infest these animals and warning them of approaching danger by their own independent reactions to it. In England starlings and sheep may form a similar association. Examples of commensal associations which confer shelter on one of the partners are those which are formed by prawns which live inside sea cucumbers and shrimps which live inside sponges. Some species of sponges have, indeed, been described as living hotels. In the loggerhead sponge, for instance, no less than 17,128 other animals, belonging to ten different species, have been found.

Another form of animal association whose origin can be traced ultimately to the need to obtain essential food substances is called *symbiosis*. This word, which literally means 'living together', signifies not only a co-operation between two different kinds of organisms which secures for both certain necessary constituents of the food, but also an association more intimate than commensalism. It amounts, in its typical form, to a physiological dependence of each partner upon the other, and frequently, but not always, it involves close physical contact between the tissues of the two partners. The physiological interdependence is, moreover, permanent. Each partner supplies to the other some element of the food without which life is not possible, and neither can, with the exception of certain phases of the life history whose function it is to find and combine with the other partner, live an independent life.

An example of symbiosis between two different kinds of animals is the association between white ants (termites) and the single-celled flagellated organisms which live in their food canals and assist these ants to digest the cellulose in the wood upon which they feed. An example of symbiosis between an animal and a plant is the association between the single-celled green plant (*Zoochlorella*), which lives between the cells of the fresh-water animal called *Hydra viridis*, a relative of the sea anemones. *Zoochlorella* produces, as other green plants do, oxygen which the *Hydra* needs and uses; the waste matter produced by the *Hydra* contains nitrogen which *Zoochlorella* requires. Examples of symbiosis between two different kinds of plants are provided by the lichens, which are associations between a fungus and the kind of single-celled plant called an alga.

If we now compare these two forms of nutritional association with parasitism, we find that parasitism, in its simplest form, is an association between one partner, called the *parasite*, which obtains, by a number of different methods, its food from the body of the other partner, which is called the *host* of the parasite. Parasitism, therefore, is, like commensalism and symbiosis, an association which is primarily based upon a need for an adequate food supply. It is more than this—much more, as the rest of this book will show; but the primary basis of it is, without doubt, the parasite's imperative need for food. The parasite, therefore, like the commensal and the symbiont, benefits from its parasitism in this respect.

But what about the other partner—the host? Does this also benefit? The answer is that it never does. The host is always injured by the parasite. Parasitism therefore differs from both commensalism and symbiosis in two fundamental particulars. First, not both the partners in it, but only one of them, the parasite, gains a food supply. Secondly, not both, but only one of the partners is benefited by it; the other, the host, always suffers some form and degree of injury. The injury done to the host may cause its death; or it may be less severe, yet severe enough to constitute what we call a disease; or it may be even less severe than this, or so slight that it causes the host no more than a temporary inconvenience; or it may, in its slightest form, be discernible only by the methods of the experienced biologist. Yet, whatever its kind or degree, the injury to the host is always there; and this injury done to the host is the most reliable feature by which we can distinguish parasitism.

Because the parasite obtains its food from the body of its host, parasitism must, like symbiosis, involve a physiological dependence of the parasite upon its host. Frequently, like symbiosis, it involves intimate contact between the tissues of the parasite and those of the host. In these two respects parasitism is more closely related to symbiosis than to commensalism. Originally, we may suppose, when parasitism first began, a non-parasitic organism penetrated, by one or other of the routes discussed in the course of this book, the body of another kind of animal. There it found food; and often it found food, such as blood, which was rich in nutritional elements and readily digestible. Thereafter, in the course of evolution, its descendants maintained the association thus established with this other kind of animal, which thus became the host, or supplier of food, for this non-parasitic animal exploiting a new and, in some respects, easier way of living. For a time the non-parasitic animal, now in process of becoming a parasite, but not yet fully committed to that mode of life, was not entirely dependent upon its host for its food supply. It could still live independently and so was, as many parasites are to-day, only a facultative and not an obligatory parasite. But later, in the course of the struggle for survival, it became more and more dependent upon its host and eventually could not live without it. In this way it became an obligatory parasite entirely dependent physiologically upon its host.

Meanwhile, however, the host did not tolerate passively its association with the parasite. It reacted to the injury done by the parasite, fighting it by methods described in Chapter 7. The struggle between host and parasite, the details of which are the major theme of this book, went on in obedience to the laws of evolution; and still, as we know, this battle is constantly being waged to-day.

There are some hosts and parasites, however, which began, in the course of evolution, to mitigate the severity of their struggles and to adapt themselves to each other. The host began to suffer less and less, the parasite to do less injury and to benefit by the host's tolerance of it; and thus, by gradual stages, there arose a state of equilibrium between the host and the parasite, a mutual adaptation and tolerance, which caused the minimum of inconvenience to the host compatible with sufficient benefit to the parasite. Thus arose, no doubt, those numerous instances of parasitism in which the host is certainly injured by the parasite, but suffers so little that its health and successful life are not compromised.

This state of mutual tolerance or equilibrium is further discussed in Chapter 7. Some biologists think that it may become very like symbiosis. It still differs fundamentally from symbiosis, because the latter is a form of co-operation for mutual benefit, while the former is still a mutual tolerance established between two antagonists. Whether, in the further course of evolution, these forms of parasitism which have reached the phase of mutual tolerance have ever been converted, by the removal of the injury of the host, into the true co-operation for mutual benefit characteristic of symbiosis is a question which must, for the present, be left aside.

The statement of it at this point serves, however, to emphasise the important fact that the physiological conception of parasitism is the only true conception of it, and to indicate the further important fact that parasitism and symbiosis are, in a physiological sense, related. Both are intimate physiological associations between different kinds of animals, or between animals and plants; both have been established primarily for the nutrition of the organisms taking part in them. The difference between them is that symbiosis benefits both partners, while parasitism benefits only the parasite.

Neither the fact that the parasite gets its food from the host nor the damage which it inflicts upon the host is, however, sufficient by itself to give us an adequate definition of the parasitic way of life. If we used only these two features for our definition, we should have to include the predatory animals among the parasites; for they harm and feed upon the animals upon which they prey. In our ponds and ditches there are carnivorous, single-celled animals which swim about in the aquatic jungles of water plants and prey upon the other single-celled animals and plants which they find there. In our gardens the brown centipede, *Lithobius forficatus*, and the carnivorous slugs belonging to the genus *Testacella*, which carry small shells upon their hinder ends and are sometimes called the gardener's friends, feed upon earthworms, slugs and other small animals, some of which do much harm to garden crops. Many birds feed upon caterpillars. Caterpillars feed upon plants. In Africa and India the lion and tiger prey upon other members of the animal communities in which they live. All these are predatory animals, and in a sense they form, like the parasites, physiological associations with the other animals from which they derive a food supply. Yet we do not for this reason call them parasites. They are parasitic only if we use the word parasite so loosely that it loses its usefulness.

Their relationship to their prey is certainly, like that of the parasite to its host, nutritional, but, unlike the parasite's relationship with its host, it always involves the death of the prey and the consumption of its body, or part of it, by the predator. So long as both predator and prey are alive, there is no physiological interplay between them and there is no organic contact at all. The association between the parasite and its host may, on the other hand, exist without the consumption of the body of the one by the other and without the death of the host. The death of the host is, indeed, greatly to the disadvantage of the parasite, and, although some parasites may destroy a part, or even the greater part, of the bodies of their hosts, the majority of parasitic associations result in a mutual adaptation between the host and the parasite which enables both to live and to propagate their species.

The parasitic relationship, moreover, usually requires an intimate contact between the tissues of the parasite and its host. The parasite needs to breathe and excrete waste material; and the host reacts against the presence of the parasite in its body and tries to get rid of it or to combat its effects. Complex physiological adaptations between the host and the parasite therefore develop, which constitute a relationship which is fundamentally different from that which exists between a predator and its prey.

There is one other feature of parasitism which also helps us to distinguish between the predator and the parasite. The predatory animal is very often, although not always, bigger and stronger than its prey. The parasite, on the other hand, is usually smaller and physically weaker than its host. If we add this feature to the other characteristics of parasitism which have just been described, we can construct the following definition of the parasite:

A parasite is a living organism which establishes a physiological association with another living organism, usually belonging to a different species and usually bigger and stronger than the parasite, which is called its host, the parasite living either on the surface of, or inside, the body of the host and inflicting upon it some degree of injury, against which the host reacts in various ways.

This definition is a bionomic definition. It is, that is to say, based upon the study of the bionomics, or way of life, of the parasite. Like all definitions of living things, it is a category imposed by the human mind upon one section only of the constantly evolving and continuously changing process which we call life. We cannot, therefore, expect that all the living things to which it refers will fit neatly into it. What has been written about commensalism, symbiosis and parasitism will have made this clear. The same observation applies, for the same reason, to classifications of parasites which are based upon nothing else than particular features of the ways in which they live.

We can, for example, classify parasitic organisms into those which are parasitic only when they suck blood or other food from their hosts. Examples of them are the mosquito and the leech, and they have been called *temporary parasites*, to distinguish them from other species which are parasitic throughout the whole or the greater part of their life histories and are therefore called *permanent parasites*. There are also the *obligatory parasites* already mentioned, which must lead a parasitic life and cannot live in any other way, and the *facultative parasites*, which can live either a parasitic or a non-parasitic life. The maggots of some species of bluebottle and greenbottle flies and also those of some of their relatives (see Chapter 9) normally live on decaying animal or vegetable matter; but if the flies into which they develop lay their eggs upon wounds or sores upon the bodies of men and animals, the maggots of these species then can become parasitic in these wounds and feed upon the matter (*pus*) in them.

In addition to the four categories just described there are several others which are *bionomic*. They are, that is to say, based upon particular features of the parasitic animal's way of life. There is, for instance, the division of parasitic animals into *ectoparasites*, which live on the surfaces only of their hosts, and *endoparasites*, which live inside their bodies. These terms are frequently used and for certain purposes they are useful; but they refer only to the position occupied by the parasitic animal and they may be misleading.

For a parasitic animal, like any other living thing, must be considered in relation to the totality of its environment. It is not sufficient to consider only the situation in which it lives. Whatever its situation in or on the host's body, it may provoke reactions of the host against it which are described in Chapters 7, 8 and 9 of this book, and these reactions may extend beyond the actual part of the host occupied by the parasitic animal. Whether the parasitic animal be ectoparasitic or endoparasitic, it may provoke immunological reactions which are made by the host acting as an organised whole, and several of the host's internal organs may take part in these reactions. When this happens, the distinction between the ectoparasite and the endoparasite has little more than a topographical value.

A sucking louse, for instance, which is certainly an ectoparasite, because it lives entirely upon the surface of its host, sucks its host's blood and tissue fluids, and may, through this contact with the host's tissue fluids, provoke reactions of the host which involve the host's internal organs.

A mosquito, which is also an ectoparasite, because it never enters the interior of its host, also sucks the host's blood, and may cause responses of the host's internal organs;

it may even introduce other parasitic animals, or micro-organisms, which can cause a disease affecting the host's whole organisation.

The mites which cause scabies of man, sheep and other animals are usually classified as ectoparasites. But some species of them, such as the species of the genus *Psoroptes*, which cause scabies of sheep, live on the surface of the host, while others, such as the species of the genus *Sarcoptes*, which cause scabies of man and some forms of mange of animals, burrow into the layers of the skin and live in these. They therefore live, in a physiological sense, actually inside the host.

Other examples could be given of ectoparasites which cannot, for reasons like these, be separated, except topographically, from endoparasites. Certainly we cannot, when we are considering the entry of parasitic animals through the host's surface, maintain a distinction like this, which, although it is useful for some descriptive purposes, has no basis in the biology of the parasitic animal concerned. For this reason—and it is a fundamentally important one—the terms ectoparasite and endoparasite, together with their respective adjectives, will be avoided as far as possible in this book.

Other categories of parasitic animals, based upon their relations with other organisms, are the *hyperparasites*, which are themselves parasitic upon other parasites, and the *brood and social parasites*, which are such an interesting feature of the communal life of ants, bees and wasps; and there are others which are mentioned elsewhere in this book. In this chapter, however, a few words must be given to: (*a*) the zoological classification of parasitic animals according to their genetic relationships with other animals; and (*b*) the division of them into (1) species which pass directly from one host to another and (2) species which cannot do this, but must pass part of their lives in one host, which is called the intermediate host, before they can pass on to another host, which is called the definitive host.

THE ZOOLOGICAL CLASSIFICATION
OF PARASITIC ANIMALS

The term zoological classification of parasitic animals is here used to mean the position which they occupy in the scheme of the relationships between animals which the zoologist bases upon their structure, development and heredity. Most of the parasitic animals belong to the groups of animals which have no backbones (Invertebrata). Very few backboned animals (Chordata) are parasitic.

One reason for this, no doubt, is the fact that the back-boned animals, with the exception of the sea-squirts and their relatives, are more highly organised and efficient than the invertebrates, so that they are well able to maintain themselves without resort to the dependence upon other animals which parasitic life imposes. They are also relatively bulky animals, with well-developed skeletons, muscles and nervous systems, which give them ample opportunities to exploit modes of life other than parasitism. Backboned animals, moreover, appeared later in history than the invertebrates, so that they have had less time in which to try out the possibilities of parasitic life and fewer opportunities to investigate it. Practically the only back-boned animals which are parasitic in the sense defined above are the few species of bats (Cheiroptera) which suck blood, and the hagfishes, which are close relations of the lampreys or 'nine-eyes'.

The question whether the young of all warm-blooded, hairy animals which suckle their young (mammals) are, while they are developing inside the womb (*uterus*) of the mother, parasitic upon the mother is an interesting one. The embryos of some invertebrate animals also develop inside the wombs of their mothers, and among them are the embryos of some animals which are parasitic, such as the embryos of the pork trichina-worm and the guinea-worm, both of which may be parasitic in man.

All embryos which thus develop in the womb of the mother abstract food from her. They do not, as a rule, do her any appreciable harm. They may, in fact, exert a beneficial effect upon her health. The human female, for example, certainly enjoys, on the whole, better health if she has children than if she does not. But the fact remains that embryos which develop inside the bodies of their mothers do exert profound effects upon the physiological processes of the mother; they are in vital physiological union with her and they may do her harm. The mammalian embryo especially may, under certain circumstances, do the mother direct or indirect injury, or may lower her resistance to the effects of other species of parasitic animals

PLATE I

Fig. 1. The liver fluke of sheep and cattle, *Fasciola hepatica*. Length 30 mm., breadth 13 mm.

Fig. 2. The common tapeworm of the sheep, *Moniezia expansa*. Length up to 600 cm. Note that the tapeworm and the fluke (fig. 1) have the flattened body characteristic of the flatworms.

Fig. 3. The large roundworm of the pig, *Ascaris lumbricoides*, not full-grown. Full-grown female 20–40 cm., full-grown male 15–30 cm. long.

Fig. 4. The large lungworm of the sheep, *Dictyocaulus filaria*. Female 5–10 cm., male 3–8 cm. long. Contrast the smooth, unsegmented bodies of the roundworms shown in figs. 3 and 4 with the ringed, segmented body of the annelid earthworm and leech (figs. 5 and 6).

PLATE I

Fig. 1

Fig. 2

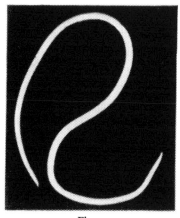

Fig. 3

Fig. 4

or to those of bacteria or viruses; or it may be the cause of diseases or injuries which kill the mother. It is possible, indeed, that, when this method of development of the young first appeared in the course of evolution, it was the cause of much more mortality among the mothers than it now is. Some species of parasitic animals, moreover, which cause serious diseases of the mother or even her death, may reach her through the placental blood circulation by means of which she nourishes the young in her womb (see Chapter 9).

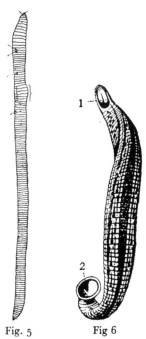

Fig. 5   Fig 6

These are all reasons which have led to the view that the mammalian embryo, at any rate, is quite definitely parasitic upon the mother in whose womb it develops. Certainly it is difficult to exclude it from the definition of the parasite given above, or to say how its way of life differs from that of the young of species of animals, such as the warble-flies, whose young (larvae) are parasitic, while their adult phases are not (see Chapter 3).

Fig. 5. An earthworm, *Lumbricus terrestris*, showing, in contrast to the unsegmented roundworm (figs. 3, 4), the ringed, segmented body. Length 9–30 cm.

Fig. 6. The medicinal leech, *Hirudo medicinalis*. Length 8–12 cm. 1. Anterior sucker round the mouth. 2. Posterior sucker

Among the animals which have no backbone the majority

16

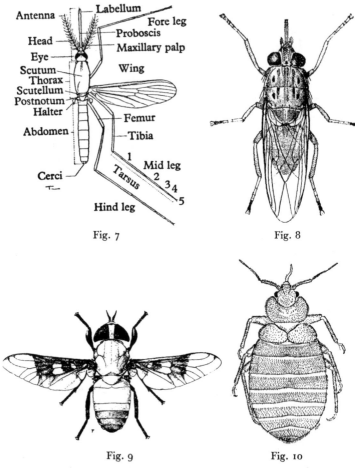

Fig. 7

Fig. 8

Fig. 9

Fig. 10

Fig. 7. Diagram of a female mosquito

Fig. 8. A tsetse fly, *Glossina longipennis*, in the resting position with folded wings. Length 10·6–11 mm.

Fig. 9. A tabanid fly (horse-fly type), *Tabanus latipes*. Length 20 mm. Note that this insect and the tsetse fly (fig. 8) and the mosquito (fig. 7) are all Diptera with two wings only

Fig. 10. The bed-bug, *Cimex lectularius*. Length 4–5 mm. A wingless insect (cf. figs. 88–9)

PLATE II

Fig. 11

Fig. 12

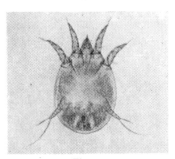

Fig. 13

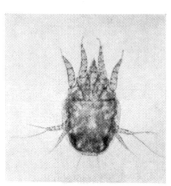

Fig. 14

Fig. 15

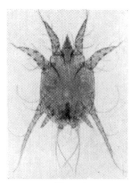

Fig. 16

of parasitic species belong to four only of the major divisions (Phyla) into which the zoologist divides these animals. These divisions are:

(a) The single-celled Protozoa (figs. 17–21), which are the most primitive of all animals. Examples of parasitic animals which belong to this phylum are the parasites which cause human malaria and the trypanosomes (fig. 18) which cause human sleeping sickness in Africa and South America.

(b) The flatworms (Platyhelminthes), a phylum which includes the flukes (fig. 1) and the tapeworms (fig. 2).

(c) The roundworms (Nemathelminthes), examples of which are the large roundworm of man and the pig (fig. 3) and the lungworms of sheep and cattle (fig. 4). A non-parasitic member of this phylum is the vinegar eelworm which is often found in vinegar barrels or even in the vinegar in our cruets.

(d) The large phylum of animals called the Arthropoda, which includes the crab and lobster section (Crustacea), the insects (Insecta) (figs. 7–10) and the scorpions, spiders, ticks (figs. 11, 12) and mites (figs. 13–16) and their relatives (Arachnida) and other non-parasitic species.

Among the other groups of the invertebrate animals there are some parasitic species, but of these the leeches

PLATE II

Fig. 11. The castor-bean tick, *Ixodes ricinus*. Length 11 mm. when engorged

Fig. 12. The castor-bean tick laying eggs

Fig. 13. The sheep-scab mite, *Psoroptes communis* var. *ovis*, six-legged larva. Length about 0·3 mm.

Fig. 14. The sheep-scab mite, nymph. Length about 0·4 mm.

Fig. 15. The sheep-scab mite, ovigerous female. Length 0·67–0·75 mm.

Fig. 16. The sheep-scab mite, adult male. Length 0·5–0·6 mm.

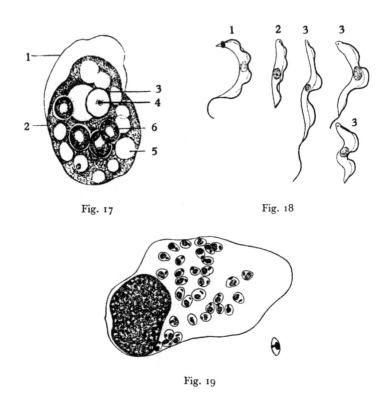

Fig. 17   Fig. 18

Fig. 19

Fig. 17. *Entamoeba histolytica* unencysted. Diameter 15–60 microns. 1, ecto-plasm; 2, endoplasm; 3, nucleus; 4, karyosome; 5, vacuoles; 6, ingested red blood cells

Fig. 18. Trypanosomes: *Trypanosoma cruzi*. Length 20 microns. 2. *T. congolense*. Length 9–18 microns. The flagellum does not extend beyond the body. 3. *T. gambiense*. Three of the different forms it takes during its life history. Length 14–33 microns

Fig. 19. *Leishmania* sp. A number are shown inside a large phagocytic cell (macrophage) of the host and one is shown outside it. Length 2–3 microns

only will be discussed in this book. The leeches, like the non-parasitic earthworms (fig. 5), belong to the phylum of ringed, segmented worms called the Annelida.

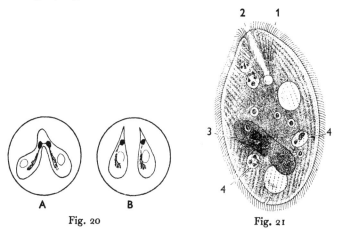

A   B

Fig. 20   Fig. 21

Fig. 20. *Babesia* (*Piroplasma*) *canis* inside the red blood corpuscles of the dog. A, division in progress; B, division completed. Length 4·5-5 microns

Fig. 21. *Balantidium coli.* Length 50-100 microns. 1, locomotive cilia; 2, mouth; 3, meganucleus; 4, food vacuoles containing partly digested fragments of the host's tissues and red blood cells

DIRECT AND INDIRECT LIFE HISTORIES

During the lives of all animals, whether they be parasitic or not, a series of phases of the animal succeed one another from the fertilised egg to the sexually mature adult, and this succession of phases is called the life history or life cycle of the animal. The life history of a butterfly, for instance, begins with the fertilised egg, which is succeeded by the caterpillar, and this turns into the chrysalis, from which the butterfly emerges.

The life histories of different kinds of animals are very varied, and zoologists have found it necessary, for a number of reasons, to give precise names to the different phases which occur during them. Thus the name *embryo* is given to the phase of the young individual so long as it remains inside the egg-shell or other protective envelope or inside the body of the mother. The name *larva* is given to a young individual which is structurally different from the adult phase after it has developed sufficiently to feed itself and to lead an independent life.

The larva of a butterfly, for instance, is the caterpillar; the larva of the house fly is the maggot; the larva of a lobster is a particular kind of larva called by zoologists a *nauplius* (cf. Chapter 5); and there are other names given to various larval phases which we shall have to use in the course of this book. Those which concern us most are given in Chapters 3 and 4.

The eggs and the larvae of all animals play, of course, an important part in the maintenance of the species and in its distribution about the world. The eggs and larvae of parasitic animals are especially important in this respect, because either the egg or one of the larval phases of the parasitic animal has to make the transition from one host to another. The success of parasitic life depends upon the successful accomplishment of this transition, so that it merits special attention. This transition may be either direct or indirect. What do these terms, direct and indirect, precisely mean? Let us consider direct transition first.

A parasitic animal which passes directly from one host to another is said to have a direct life history. It lives in host *A* and produces in host *A*, by sexual processes, either fertilised eggs or larvae which leave host A and become parasitic in host B.

Thus the sheep stomachworm (figs. 65 and 110) whose life history (fig. 22) is described in Chapter 3, produces, in the food canal of the sheep (host A) fertilised eggs which are passed out in the sheep's droppings. From these eggs emerge larvae which live a non-parasitic life on the pastures until they develop into a larval phase which is eaten by another sheep (host B) when this sheep crops the grass. Until this particular phase has been reached the larva cannot infect the second sheep.

Inside this second sheep this larval phase grows up into the adult phase again.

When the life cycle of a parasitic animal follows this general plan it is said to be direct and the parasitic animal is called a *monoxenous* species. Life cycles of this kind may be represented by the simple diagram

$$A \rightarrow B.$$

In contrast to life cycles which are direct, there are life cycles of parasitic animals which are indirect. These species cannot pass directly from host A to host B.

Before they can enter host B, the larvae which hatch from their eggs must become parasitic and develop in another kind of animal, which is usually, but not always, an animal biologically very different from host A. Frequently, when host A is a back-boned animal, this other host is an animal without a back-bone, such as a blood-sucking insect, tick or leech or an aquatic, air-breathing snail.

This other host may be called host m. In it some of the larval phases of the parasitic animal must grow and develop until a larval phase is reached which leaves host m and passes into host *B*. Only this particular larval phase which leaves host *m* is able to infect host *B*.

A life cycle of this type may be represented by means of the simple diagram $A \to m \to B.$

The host (host *m*) inside whose body the larval phase of the parasitic animal must develop before it can infect hosts *A* or *B* is called the *intermediate host*. The host inside whose body the sexual processes of the parasitic animal produce the fertilised eggs (host *A* or *B*) is called the *definitive host*. When the life cycle follows this plan it is said to be indirect and the parasitic animal is called a *heteroxenous* species.

An example of an indirect life history is the life history (fig. 37) of the liver fluke of cattle and sheep described in Chapter 4. The adult phases of this fluke live in the bile canals of the sheep's liver and produce eggs which pass out of the sheep in its droppings. The sheep is thus one of the definitive hosts of this species (host *A*). From each of these eggs hatches out a larval phase which must enter and develop inside an amphibious, air-breathing snail, which is therefore the intermediate host (host *m*). As a result of this period of development inside the body of the snail a larval phase called a cercaria is produced and this leaves the snail to enter another sheep (host *B*) and to become, in that sheep's liver, the adult phase again.

The examples just given of the direct and indirect life histories omit, for the sake of simplicity, an important additional feature which must be explained.

The stomachworm of sheep, which has a direct life history, is not confined to sheep; it can live parasitically in goats, cattle and some other animals that chew the cud. The letters *A* and *B* used in the diagram of the direct life history just given may therefore refer to all these other hosts. The human hookworm, *Ancylostoma duodenale*, on the other hand, is practically confined to man. In this instance, therefore, the letters *A* and *B* refer to man only.

Among species with indirect life histories, the liver fluke (*Fasciola hepatica*) can use as its definitive hosts, not only sheep and cattle, but also pigs, rabbits, hares, dogs, cats, elephants, man and other animals. Although in Britain it can use only one species of snail (*Limnaea truncatula*) as its intermediate host, in other countries it can use other snails. In this instance, therefore, the letters *A* and *B* may represent several different kinds of warm-blooded animals, and the letter *m* may, in different countries, represent different kinds of aquatic or amphibious snails, belonging to more than one genus. The human malarial parasites, on the other hand, can use only man as their intermediate host, and the mosquitoes which they can use as definitive hosts all belong to the genus *Anopheles*. In a diagram of their life histories, the letter *m* would refer to man only, and the letters *A* and *B* would refer to Anopheline mosquitoes only.

It will be evident, therefore, that some species of parasitic animals are restricted to a few hosts, whether these be definitive, intermediate or, if the life history is direct, neither; others are less restricted. When a species is restricted to only one species of host, it is said to be *specific* to that species of host. The study of this *host-specificity* is biologically very interesting, and a knowledge of the host-specificity of the species of parasitic animals which cause diseases of man and domesticated animals greatly helps to control or eradicate these diseases.

CHAPTER 2

HOW THE PARASITIC ANIMAL
MAKES CONTACT WITH THE HOST
AND MOVES ABOUT INSIDE IT

Before parasitic life can begin, the parasitic animal must make contact with the tissues of its host; and, once this contact has been made, it must be maintained. Almost every kind of animal tissue suffers from the effects of parasitic animals and there are few, if any, kinds of animals which do not, at some time during their lives, act as the hosts of one kind of parasitic animal or another. All these different kinds of hosts show considerable differences of structure, physiology and mode of life. The insect is quite different from the backboned man and his dog; the earthworm in our gardens and the animals that live in fresh water and the sea are different from reptiles that live on land or birds that live in the air. The host, moreover, is an animal, and animals are active creatures, with which it is not easy to make contacts. The primary problem of the parasitic animal, therefore, the problem of making contact with the host, would seem at first sight to be not an easy one.

To look at the problem in this way, however, is to put the cart before the horse. Each parasitic animal has been derived, in the course of evolution, from a non-parasitic ancestor, and this ancestor was an animal which was, when parasitic life began, normally associated with the host in which its descendants became parasitic. The parasitic glochidial larva of the swan mussel, for instance, uses as its host a fish which normally swims past the swan mussel; the human malarial parasites were doubtless originally parasitic only in the mosquitoes which later transmitted them to man when the female mosquitoes fed upon human blood. Originally, no doubt, the parasitic animal was confined to the host in which it first became parasitic; but later many species learned, as we know, to become parasitic in other hosts as well. Usually, however, we find that these additional hosts are animals which live in more or less close association with the first host, and in some instances we know that the transition from one host to another depends entirely, as the transition of the pork trichina-worm does, for instance, upon this close association.

If we look at the matter in this way, we see that the various methods by which the parasitic animals of to-day enter and leave their hosts have gradually evolved. They are the methods which have proved to be, in the struggle for existence, the ones that are most favourable to the survival of the parasitic animal. What are these methods and how are they related to the movements and behaviour of the parasitic animal when it has made contact with its host?

The problem of making contact with the host is, perhaps, easier for animals which are parasitic only temporarily upon the surfaces of their hosts and do not enter their internal tissues. The mosquito, the leech and the blood-sucking bat have only to reach and stay upon the host's surface long enough to enable them to take the meal of blood they require. The species which live permanently on the surfaces of their hosts have a more difficult problem, but it is not so difficult as that of the species which not only have to reach their hosts, but must also succeed in entering their bodies and in adapting themselves to life inside those bodies.

One of the easiest ways of entering the interior of the host is to lie in wait for it, either in its food or drink, and so to enter it through its mouth. This is probably the reason why more species of parasitic animals enter their hosts through their mouths than in any other way.

Other possible ways into the interior of an animal are through the other natural openings on the surface of its body, namely, the ways into its breathing organs or the openings which let out the products of its reproductive organs or those of its kidneys or of the organs which correspond to the kidneys. Yet another method of entering is by actual penetration of the surface of the host's body.

All these possible ways into the host are indicated in the simple diagrams (figs. 22–24, 33–39, 54, 55) which illustrate the life histories of parasitic animals discussed in Chapters 3 and 4. These diagrams represent schematically certain features of the bodies of one kind of host only, namely, the mammal; but similar diagrams could be constructed of the bodies of other kinds of animals which act as hosts. In the diagrams here used the body of the host is represented as a circle, and the stippling inside each circle indicates the tissues of the host's body. The line enclosing the circle represents the surface layers of the host. Through the host's body, from its mouth to its vent (*anus*), passes a tube, which represents the food canal. We know that the food canal of mammals consists of the gullet, stomach and the long coils of the intestines, but it is represented here, for the sake of simplicity, as a straight tube. Certain organs through which many species of parasitic animals must pass are sketched in. These organs are the liver, which has arisen, in the course of evolution, as an outgrowth of the food canal; the lungs, which have also arisen in this way; and the heart. The right and left lungs are, for the sake of simplicity, both drawn in these diagrams on one side of the food canal.

But why, the reader may ask, are these organs, and no others, put into the diagrams? The answer is that many species of parasitic animals pass through these organs, the liver, the heart and the lungs, and they pass through them because they all tend to be guided to these organs by the natural course of the circulation of mammalian blood.

When a new host is entered, or, as the parasitologist puts it, when it is infected, the phase of the parasitic animal which enters the new host is called the *infective phase* of the parasitic animal. This infective phase may be either the egg, the larva which succeeds the egg, or some other larval phase which develops later in the life history. Some infective phases are active. Either they have definite locomotive organs, or they move by wriggling, boring and other movements of their whole bodies. Other infective phases, such as the egg, are passive and quite unable to move by their own efforts. Passive infective phases cannot, of course, influence their entry into the host. They must wait until the host itself takes them in. Usually they are taken in with the host's food and drink, so that they normally belong to species which enter the host through its mouth. Active infective phases, on the other hand, do by their own normal activities influence their entry into the host. If they normally enter the host through its mouth, they can, by their own movements, place themselves upon or in its food or in its drink. If they enter it through its outer surface, they can bore their way through this. After they have entered the host both active and passive infective phases are carried about in the host's blood or other tissue fluids or migrate by their own efforts to the parts of the host's body in which their adult phases normally live. The routes which they follow can be classified according to the portal of entry used by the infective phase.

A. Entry by the Mouth

When the infective phase enters by this route, it finds itself in the food canal. It then takes one of the following main routes:

(A) *It remains in the food canal*

It grows up to the adult phase in this situation. The parasitic animal does not, at any stage of its life history, go beyond the food canal.

There are two chief variants of this course:

(*a*) The infective phase remains in the contents of the food canal, growing up in these into the adult phase, which produces fertilised eggs. The eggs are then passed out of the host through the anus with the excreta. Fig. 22, of the life history of the stomachworm of sheep and cattle, illustrates this simple traffic route. It may be stated thus:

External world—mouth—contents of the food canal— anus—external world.

(*b*) The infective phase, after reaching the contents of the food canal, burrows into the wall of this canal and goes through part of its development there. Later, before it has become the adult phase, it passes back into the contents of the food canal to become adult in these. The fertilised eggs are then passed out of the host through the anus with the excreta. This traffic route can be stated thus:

External world—mouth—contents of the food canal—wall of the food canal—contents of the food canal—anus— external world.

The dwarf tapeworm, *Hymenolepis nana* (see Chapter 4), uses this route.

(B) *It passes beyond the food canal.*

This happens when the infective phase of the parasitic animal, after entry through the mouth, burrows into the wall of the food canal or is deposited in this by its parent, but does not stop there. It may enter either:

(1) The blood in the wall of the food canal. Larvae which enter this must pass on to the liver;

(2) The lymph-stream in the wall of the food canal or in the lymph glands near this. Larvae which enter the lymph need not pass through the liver.

(1) *It enters the blood in the wall of the food canal.* This blood is collecting the products of the digestion of the food. It is contained in the branches of the *hepatic portal vein*, an important vein which, as its name implies, carries, or is the porter of, the blood from the stomach and intestines to the liver. Any larvae of parasitic animals which get into the blood in this vein will, therefore, also be carried to the liver. This is the physiological reason why the liver is one of the main traffic centres used by parasitic animals which enter their hosts through the mouth. Fig. 37 shows that a variant of this route may occur, the larvae reaching the liver by actively migrating across from the wall of the food canal to the liver, without entering the blood stream. Sometimes, as fig. 35 indicates, the larvae may use both of these variants.

When the larva of the parasitic animal thus reaches the liver it follows one of two main courses:

(a) *It remains in the liver.* Here it grows up into the adult phase, which does not go beyond the liver, but produces fertilised eggs in it. These eggs pass back into the food canal by way of the bile ducts, which are discharging into this canal the bile which helps in the digestion of the fats of the food. The eggs then pass out of the host by way of the anus with the excreta. This traffic route is followed by the liver fluke of sheep and cattle. It is illustrated by fig. 37 and it may be stated thus:

External world—mouth—wall of the food canal—hepatic portal vein—liver—bile duct—food canal—anus—

external world.

(b) *It goes beyond the liver.* This happens usually because the larvae, which have reached the liver by way of the hepatic portal vein, are carried to the liver, in the venous blood which is going to the lungs to be aerated, so that the larvae also are carried to the lungs.

They stay in the blood in the lungs only until they have reached a certain stage of development. They then break out of the small blood vessels (*capillaries*) in which they have developed into the air sacs and wriggle up the air passages to the upper end of the trachea. Reaching the junction of the windpipe and the gullet, they are swallowed while they are still in the larval phase, and so pass back into the food canal from which they started. They do not grow up into the adult phase until they reach the food canal, in which they produce eggs, which pass out of the host by the anus with the excreta.

This traffic route is followed by the large roundworm,

Ascaris lumbricoides, of man and the pig. It is illustrated by fig. 35. It may be stated thus:

External world—mouth—food canal—hepatic portal vein
—liver—right side of the heart—lungs—air passages—
gullet—food canal—anus—external world.

(2) *It enters the lymph stream in the wall of the food canal or in the lymph glands near this.*

The larvae of the lungworms (see Chapter 9) and those of the pork trichina-worm (see Chapter 3) follow this route. The larvae of the lungworms burrow through the intestinal wall to the mesenteric lymph glands and from these they travel, by way of a lymph tube called the *thoracic duct* (see Chapter 3), to the large vein in the left-hand side of the neck which collects venous blood from the head, neck and arm of that side and conducts it to the heart. With this blood the larvae go to the lungs and they grow up in these organs or in the air passages into the adult phases. The adults are parasitic here and produce eggs or larvae, which reach the top of the windpipe (*trachea*), but do not, as we might expect, pass out of the host by the mouth or nose. A few of them may leave the host in this way, but most of them are swallowed into the food canal again, so that they leave the host by the anus with the excreta.

A similar route is followed by the larvae of the pork trichina-worm, but the larvae of this worm are deposited by their parents in the lymph spaces in the wall of the host's food canal. They then travel, by way of the thoracic duct, to the venous blood stream, by which they are carried to the right side of the heart. (Some of them may enter the blood in the hepatic portal vein and reach the heart by way of the liver.) From the heart they go to the lungs, but, unlike the larvae of the lungworms just described, they do not stay in these organs. They leave them in the oxygenated blood, which takes them back to the heart, but this time to the left side of it, from which they are distributed with the arterial blood all over the host's body. They may, therefore, reach any of the host's organs. Only those, however, which settle in certain of the voluntary muscles can develop further. In these muscles they do not grow up into the adult phase. They remain in them, still in the larval phase, until the muscles in which they lie are eaten by a new host. Reaching the food canal of this new host they repeat the same cycle. This traffic route, therefore, differs from the others described in two important respects:

(1) It involves for its completion two separate hosts, neither of which is an intermediate host.

(2) It excludes the parasitic animal entirely from the external world.

It can be stated thus:

Wall of food canal—general blood stream—voluntary muscles—mouth of new host—wall of food canal of the new host.

B. Entry through the Vent (*Anus*)

It is doubtful whether any species of parasitic animal now uses this way into its host; but it is possible that some species, especially those parasitic in the food canal, originally entered by this opening and later abandoned it for another way in. The larvae of the human threadworm can, however, enter man through the anus (see Chapter 9).

C. Entry through the Openings of the Respiratory, Renal and Reproductive Organs

It is an interesting fact that, although many different kinds of parasitic animals live in all these organs, the infective phases of most of them enter their hosts either through the outer surface or through the mouth. This fact is further discussed in Chapter 9, in which the species which do enter by the openings of the respiratory, renal and reproductive organs are also considered.

D. Entry through the Surface of the Host

Parasitic animals which enter their hosts by this route either (a) stay in the substance of the skin, or (b) pass beyond it.

(a) *They remain in the skin.*

Examples of species which do this are the mites, described in Chapter 5, which cause scabies of man and animals and the infective larvae of certain flukes and roundworms, when these get into the skin of hosts in which they cannot grow to maturity (see Chapter 9).

(b) *They pass beyond the skin*

Species which do this either

(1) penetrate the host's surface by their own efforts; or

(2) are introduced through it by some other agency, which may be: (a) The female adult which produces the infective phase, but is not itself parasitic. A female insect, for instance, may pierce the outer surface of another animal and deposit in that animal a larva which is parasitic.

Examples of this method of entry are considered in Chapter 9. (*b*) A blood-sucking animal, such as a mosquito, tsetse fly, tick or leech, which is itself also parasitic on the same host and sucks its blood (see Chapter 8).

Examples of species whose infective phases penetrate the host's surface by their own efforts are the warble-fly of cattle and the human hookworms. The young larva of the warble-fly hatches during the summer from eggs attached to the hairs of cattle and bores its way through the skin to enter the tissues beneath. It then burrows about in these tissues during the autumn and early winter, and finally turns up in the late winter or early spring under the skin alongside the backbone. Here it feeds and grows, until it is ready to become a pupa. It then falls to the ground, pupates there and eventually gives rise to the adult fly.

This traffic route is illustrated in fig. 36. It may be stated thus:

External world—hairs on the skin—tissues under the skin— tissues of the body—tissues under the skin of the back— external world.

Other infective phases which penetrate the host's surface by their own efforts are the active larvae of the hookworms just mentioned, which bore through the host's surface and reach the venous blood stream. By this they are carried to the right side of the heart. They do not pass first to the liver, because the blood leaving the surface of the host does not necessarily pass through this organ. From the heart they pass in the blood to the blood vessels in the lungs. Like the larvae of *Ascaris lumbricoides*, they break out of these blood vessels into the air sacs, wriggle up to the top of the windpipe, pass over into the gullet and are swallowed into the food canal, where they grow up into the adult phase. The eggs laid by the female adult then pass out of the host by the anus with the excreta, the infective phase being non-parasitic in the external world.

This traffic route is illustrated in fig. 34. It can be stated thus:

External world—penetration of the host's surface—blood stream—heart—lungs—windpipe—gullet—food canal—
external world.

Alternatively, the infective phases of species which use this route are also able to enter the host through its mouth. When they do this they follow the same route, except that they pass from the food canal by the blood stream to the liver before they reach the heart. They follow, that is to say, the route used by the larvae of *Ascaris lumbricoides*.

When the infective phase is introduced through the host's surface by a blood-sucking animal, the host whose surface is penetrated is either the intermediate host of the parasitic animal, the blood-sucking animal being its definitive host, or vice versa. The mosquito, for instance, is the definitive host of the human malarial parasite, while man is its intermediate host. The routes followed in various mammalian hosts by parasitic animals introduced into them in this way vary considerably. Some, such as the malarial parasites of man, are parasitic first in the liver and then in the red blood cells; others enter the blood but are distributed by it to other organs in which they become parasitic; others leave the blood and pass into the lymph and the channels and glands which control the movements of this fluid. But no useful purpose would be served at this stage by following out the routes used by parasitic animals introduced into the host in this manner. They are described elsewhere in this book and need not be further considered here.

ROUTES OF EXIT FROM THE HOST'S BODY

The maintenance of the species of any kind of animal always takes precedence over the maintenance of the life of the individual of the species, and to this rule the parasitic animal must conform. It is therefore necessary that some phase of it shall leave the host and infect new hosts. How does a parasitic animal leave its host? In general it leaves it either through the natural openings upon the surface of the host's body or through the surface layers of the host, or by transit into the egg, larva or other young of the mother while these are still in the mother's body.

It is an important fact that the diseases which some species of parasitic animals cause are due to the damage inflicted upon the host when the eggs or larvae of the parasitic animal are leaving the host. The human disease caused by the guinea-worm, for instance, depends upon the method adopted by the female roundworm of this species for the liberation through the skin of man of the living larvae which she produces. The serious effects of the human blood flukes are the result of the irritation caused by the eggs of these species and by their passage through the walls of the bladder or lower bowel. These consequences of the exit of eggs or larvae from the host are discussed in Chapter 9.

CHAPTER 3

REPRESENTATIVE LIFE HISTORIES, I

In the preceding chapter an outline was given of the routes of entry into the host and exit from it and of some of the traffic routes which parasitic animals use inside the bodies of mammalian hosts. In this chapter and the next one these routes will be illustrated by a series of life histories of particular species of parasitic animals. The life histories chosen for description have been selected for three reasons. First, they illustrate the main traffic routes just described; secondly, they are, with one or two exceptions, the life histories of species which cause serious diseases of man and his domesticated animals, so that the reader will, as he considers them, learn something of these diseases and of the manner in which they are acquired; thirdly, they illustrate the risks and difficulties encountered by the parasitic animal and some of the adaptations which it makes in order to counter these risks. The life histories selected are arranged in two groups. This chapter includes only life histories of species which do not use an intermediate host, while Chapter 4 includes only species in whose life histories an intermediate host must take part.

Each chapter begins with species which live part of their life histories non-parasitically outside their hosts and ends with species which have entirely abandoned non-parasitic life. The two chapters together therefore illustrate the gradual abandonment of the external world by parasitic animals whose life histories are direct or indirect respectively and their complete commitment to parasitic life.

This arrangement of the life histories described must not, however, be regarded as an evolutionary series. It does not necessarily indicate the way in which complete parasitic life has been evolved. It is merely an artificial arrangement of the material used, which is intended to help the reader. If he finds, at this early stage, that the details of these life histories are tedious or too complex, he may prefer to read through this and the next chapter quickly and to return to them from time to time during his reading of the rest of the book.

SERIES A. SPECIES WHOSE TRANSMISSION FROM HOST TO HOST IS DIRECT

Type IA (see figs. 22, 65, 110): *Haemonchus contortus*, the large stomachworm of sheep.

The adult stage of this species of roundworm is parasitic in the fourth or true stomach (*abomasum*) of sheep, cattle and some other animals which chew the cud. The female worm is 18–30 mm. long and 0·5 mm. in diameter; the male is 10–20 mm. long and proportionately more slender. Both cause inflammation of the lining of the stomach (gastritis) of the host. Sheep and cattle suffering from this gastritis lose condition, fail to thrive and may die. This worm can, therefore, inflict severe losses on the farmer.

Its life history is relatively simple and is a good example of a kind of life history during which the younger larval phases are not parasitic, while the older larval phases and the adult phase are. It may be compared with the life histories of species, such as the warble-flies and the rain-worm, considered below, which reverse this arrangement, their younger larval phases being parasitic, while the later phases and the adults are not.

The adult male and female sheep stomachworms live, in the sheep's stomach, in an acid environment, for without the hydrochloric acid secreted by the glands of the stomach

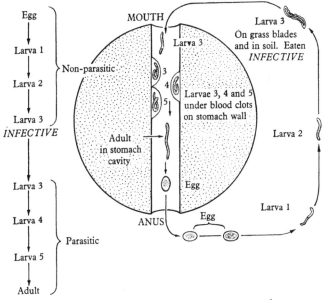

Fig. 22. Diagram of the life history of the large stomachworm of sheep, *Haemonchus contortus*, Type I A

wall, the digestive ferment, pepsin, which these glands also secrete, cannot act. The female worms lay some 5000–10,000 eggs a day for many weeks, and these eggs pass down the food canal of the sheep and out with its droppings on to the pastures. Each egg is about 77 microns long by about 45 microns broad, a micron being one-thousandth of a millimetre or one twenty-five thousandth of an inch.

On the pastures, if the climatic conditions are favourable, the contents of each egg divide repeatedly until a small *first larva* is formed. This first larva hatches out of the egg and feeds upon the bacteria which it finds in the host's droppings and elsewhere in its surroundings. When it is hatched it is about 0·4 mm. long, and its life is naturally influenced by the physical and chemical factors in its environment.

Countless numbers of these first larvae must be killed by drought, frost and similar influences; but their powers of survival are considerable, and they obtain shelter from climatic rigours inside the droppings of the sheep in which the eggs are laid, underneath the blades of grass and the leaves of clover and other pasture vegetation, or in the surface layers of the soil. The droppings of some animals, such as those of the cow, make natural incubators, which favour the development of larvae like those which we are considering. The droppings acquire a crust which helps to keep up a favourable and more or less constant temperature inside them; they teem with the bacteria upon which the larvae feed; they are often aerated by the larvae of dung beetles which bore holes through them and admit the air.

After a period of growth the first larva of the sheep stomachworm moults its skin and becomes the *second larva*, which also feeds upon bacteria and grows. This second larva, when it is full-grown, also moults its skin and becomes the *third larva*. The skin of the second larva, however, is not cast off. It is kept by the third larva as a loose sheath, which probably gives the third larva some protection against climatic conditions.

It is this third larva, with its loose sheath, which infects the sheep. It is the only phase of this species that can do this and for this reason it is called the *infective larva*. If the sheep, as it grazes on the herbage of the pasture, eats with the herbage the first or the second larva, neither of these phases can survive inside the food canal of the sheep. The egg, if it happens to be swallowed, will usually succeed in passing again through the sheep's food canal unharmed.

The fact that the third larva is the only larval phase that can infect the sheep is an important biological detail upon which is based the practice of moving lambs and sheep from one piece of pasture or folding to another. The infective larva is not developed, under normal conditions, until about three days after the egg has left the sheep. When climatic conditions are unfavourable—when, for instance, the weather is too cold or too hot and dry—a longer period is required. For this reason, sheep cannot infect themselves with infective larvae derived from eggs passed out with their own excreta until at least three days have elapsed If, therefore, sheep are moved to a fresh, uncontaminated area of pasture every fourth day, or even once a week, the chances of their reinfection from their own excreta are minimised. Some shepherds evidently knew this long before the biological reason for it was known, because there is a saying, attributed to Yorkshire shepherds: 'Never let the sheep hear the church bells twice from the same piece of land.'

The third larva inside its loose sheath is able to survive on the pastures for a considerable period of time, especially if its surroundings are moist and warm. It lives on the herbage of the pastures, on the surface of the soil, or in its uppermost layers and can travel up the stems and leaves of the pasture plants, especially when these are wet with dew or rain and when the light is dim in the evening or at dawn. While it is on the pasture plant, the third larva is eaten by the grazing animal and thus infects it.

In the mouth of the host, or in the paunch (*rumen*), or in the true stomach (*abomasum*), it emerges from its loose sheath and begins its parasitic life. It settles down in the stomach and causes bleeding from its walls. Underneath the blood clots which cover these bleeding points, it grows, and moults its skin to become the *fourth larva*. This grows and moults its skin to become the *fifth larva*, which then grows up to become the adult. When the adults are sexually mature, they then begin to lay eggs. Under favourable conditions the sheep stomachworm will be laying eggs some 3–4 weeks after the host has swallowed the infective larvae.

This kind of life history is thus divided into non-parasitic phases, namely, the egg and the first three larval stages, and parasitic phases which include the rest of the life history. An intermediate host is not required and infection of the host can occur only through the mouth. It is one of the simplest of all the life histories of parasitic animals and possibly represents one of the earliest methods by which animals became parasitic.

Parasitic in the food canals of sheep, goats, cattle and other animals which chew the cud, and also in the food canals of rodents and carnivores, there are many species whose life histories follow this plan. Not all of these species are, however, parasitic in the stomach. Many are parasitic in the small and large intestine, and not infrequently sheep and cattle suffer at the same time from gastritis caused by species living in the fourth stomach and from inflammation of the intestine (*enteritis*) caused by species living in the intestines.

Type I B (fig. 23): *Ascaridia galli*, the large intestinal round-worm parasitic in the small intestine of the fowl and other birds.

Although the non-parasitic larvae of the sheep stomach-worm are protected to some extent at least by the loose cast

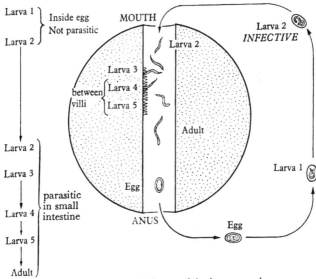

Fig. 23. Diagram of the life history of the large roundworm of poultry, *Ascaridia galli*, Type I B

skin of the second larva which encloses them, this sheath does not effectively protect them from cold, heat, drying and other injurious influences to which they are exposed in their environment outside the host: and the earlier larval stages are even more vulnerable to these. The larvae of the large roundworm of poultry, on the other hand, have a more efficient protection from the physical agencies in their environment. They develop inside the egg-shell and they do not hatch out of the egg until they have entered the new host. The infective phase is therefore an *infective egg* and not an infective larva. The infective egg is entirely passive and must rely upon being swallowed by another host. The parasitic animal is therefore unable to make any effort of its own to infect the host.

When the infective egg is swallowed, the life history of the large roundworm of poultry is similar to that of the sheep stomachworm. The main differences are that the infective larva inside the egg of the large roundworm of poultry is the second larva, not the third, and the older larvae and adults live in the intestine instead of the stomach. When the egg is swallowed, this second larva hatches out of it in the intestine of the bird host. For 9 days or so it lives free in the lumen of the intestine, moulting its skin on about the fifth day to become the third larva. The third larva lives free in the intestine for about 4 days more; but, on about the ninth day after the eggs were swallowed, it buries its head in the pits between the finger-like processes (*villi*) of the intestinal wall, which increase the area over which food can be absorbed from this part of the intestine. The rest of its body projects freely into the intestinal lumen. Lightly attached in this manner, the third larva feeds upon the lining of the intestine, and may cause bleeding as it does so. About twelve days after the eggs were swallowed, the third moult occurs. The third larva then becomes the fourth larva, which is also attached to the intestinal wall. About 19 or 20 days after the eggs were swallowed the fourth and last moult converts the fourth larva into the fifth larva, and about 7 weeks after the eggs were swallowed this becomes, without any further moults, the young sexually mature adult worm. The young adult leaves the intestinal wall to live freely in the lumen of the intestine for the rest of its life. The mature female is 72–116 mm. and the male 50–76 mm. long.

Some of the third and fourth larvae may burrow into the intestinal wall and get into the blood vessels which are taking blood from the intestine to the liver. By this blood they may be carried to the liver and thence to the lungs, from which they get back into the intestine by crawling up the air passages to be swallowed into the intestine. These larvae thus follow the route taken by the larva of their relative *Ascaris lumbricoides* described below. It is, however, unusual for the larvae of *Ascaridia galli* to take this route. Usually they develop entirely inside the intestine.

Type IC (figs. 24, 25–28): *Eimeria caviae.*

This species is parasitic in the part of the large intestine of the guinea-pig which is called the colon—the part which continues the food canal from the small intestine to the rectum, which ends at the anus. It is a single-celled species belonging to the group of animals called the Protozoa and to the section of these which is called the Coccidia. Relatives of it cause the disease of poultry and other birds called coccidiosis. The younger phases of the life history of this species are, like those of *Ascaris lumbricoides*, protected from injurious influences in the environment outside the host by an envelope which resists these influences; and its powers of resistance are remarkable.

The phase of the life history of *Eimeria caviae* which is passed out of the guinea-pig into the outside world is called the *oocyst* (figs. 27, 28). Oocysts are, as a rule, oval or spherical and are smaller than the eggs of parasitic worms.

The oocysts of *E. caviae* (fig. 27), for instance, are about 19 microns long and 16·5 microns broad. When it is first formed, the oocyst corresponds to the fertilised egg of other animals. Just as development must occur in the fertilised egg of the sheep stomachworm to produce the first larva,

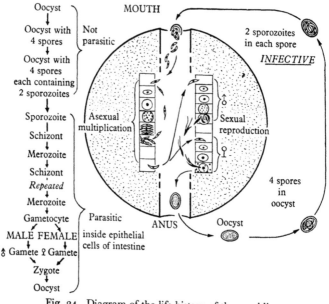

Oocyst
↓
Oocyst with 4 spores   Not parasitic
↓
Oocyst with 4 spores each containing 2 sporozoites

MOUTH

2 sporozoites in each spore

INFECTIVE

Sporozoite
|
Schizont
↓
Merozoite
↓
Schizont
Repeated
↓
Merozoite
↓
Gametocyte
↗   ↖
MALE FEMALE
↓   ↓
♂ Gamete ♀ Gamete
↘   ↙
Zygote
↓
Oocyst

Asexual multiplication

Sexual reproduction

4 spores in oocyst

Parasitic inside epithelial cells of intestine

ANUS   Oocyst

Fig. 24. Diagram of the life history of the coccidian
Eimeria caviae, Type I C

which must then develop into an infective larva before it can infect a new host, so development of the fertilised egg inside the oocyst of *E. caviae* must occur before it can infect a new host.

The contents of the oocyst divide up to form four small masses of protoplasm, called *sporoblasts*. Each sporoblast secretes around itself its own protective covering and, when it is thus enclosed, it is called a *spore*. Each oocyst of *E. caviae* therefore contains four spores. The protoplasm inside each of these spores then divides into two minute, sausage-shaped individuals which are called *sporozoites*. The production in each oocyst of four spores each containing two sporozoites is characteristic of all the species of the genus *Eimeria*. It is these sporozoites which infect a new host, and no other phase of the life history can do so. The oocyst is, for this reason, not infective until the sporozoites have developed inside it.

The process by which spores and sporozoites are formed is called *sporulation* and the infective oocyst with its contained sporozoites is called a sporulated or 'ripe' oocyst. The rate at which sporulation occurs depends upon certain factors in the environment of the oocyst. Generally speaking, warmth and moisture favour it, while cold and drying retard or prevent it. The oocysts of some species of coccidia, such as *E. tenella* (fig. 28), which causes the most rapidly fatal kind of coccidiosis of chickens, sporulate at room temperature in 48 hours. Those of other species in other hosts may, however, take much longer than this. Thus the oocysts of *E. caviae* of the guinea-pig require 5–11 days for their sporulation.

When an infective oocyst is swallowed by a new host, the two resistant membranes of which its walls are composed are altered by the gastric and pancreatic digestive juices of the host and the sporozoites are liberated from the spores. The oocysts of some species of Coccidia are provided with a small hole, called a *micropyle*, closed with a plug of material which is dissolved by the host's digestive juices, and through this hole the sporozoites emerge. The oocysts of other species, however, among which is *E. caviae*, are not provided with a micropyle, and the sporozoites cannot emerge until the walls of the oocyst break down under the action of the host's digestive juices.

When it is set free in the duodenum, the sporozoite finds its way to one of the cells of the epithelium lining the duodenal wall and bores its way into it. Inside this cell it settles down to feed upon the contents of the cell, and there it grows until finally most or all of the contents of the cell have been consumed and the cell has been destroyed. This feeding and growing phase is called the *trophozoite*.

When it is full-grown, the trophozoite divides up into 12–32 sausage-shaped individuals, each of which is 6–16 microns long. These are called *merozoites* (fig. 25). The structure of a merozoite is very similar to that of a sporozoite.

This process by which the trophozoite splits up into a number of merozoites is called *schizogony*. It is a process of multiple division by means of which the number of individuals of the species of parasitic animal is increased inside the host, and it is effected without a sexual process. The full-grown trophozoite which undergoes schizogony is called a *schizont* and, because its shape is usually oval and because its division is longitudinal, the merozoites which are formed are usually arranged much as the quarters of an orange are arranged, or their appearance may be compared with that of a bundle of closely packed, minute cigars.

Usually the number of merozoites produced when schizogony occurs is some multiple of two, and each species of coccidian usually produces a fairly constant number of them. Various factors, among which the resistance of the host is an important one, may influence the number of merozoites produced. *E. caviae* normally produces from twelve to thirty-two merozoites.

Although the structural differences between the merozoites and the sporozoites are minute, the methods by which they are formed are fundamentally different. The merozoite is the result of an asexual process of multiple division, which does not require the union of the sexes, while the sporozoite is the result of a process of division which occurs only after a union of the sexes has been effected. The functions of the merozoite and the sporozoite are also very different. The function of the sporozoite is the infection of a new host; the function of the merozoite is the infection of more of the host's tissue cells. The multiplication of the number of individuals inside the host by the asexual process which produces the merozoites may be compared to the asexual multiplication of the numbers of bacteria by division of their bodies. It means that infection with a single sporozoite may result in the presence in the host of many asexually produced individuals, whereas infection by a single infective larva of a roundworm results in the presence in the host of only one adult roundworm. The importance of this fact is further discussed in Chapters 6 and 7.

When they are fully formed, the merozoites make their way out of the remains of the cell in which they have been produced and, entering the contents of the food canal, they bore their way into other cells of the lining of this canal. Inside these cells they grow up into a second generation of schizonts, which produce a second crop of merozoites. When this second crop escapes from the cells in which it has been produced, it may infect yet more cells and may grow up in these to produce a third generation of schizonts and merozoites. Very large numbers of the cells lining the food canal may thus be destroyed and the nutrition of the host may therefore suffer considerably. Some species of Coccidia, in fact, penetrate deeper than the cells lining the food canal and may then cause severe bleeding, which increases the effects which the host may suffer. *E. tenella*, for instance, may cause in this way the deaths of many young chickens. Eventually, however, the impulse to produce merozoites spends itself. The asexual multiplication of individuals ceases and is succeeded by a sexual process.

This sexual process begins when some merozoites grow up inside their host cells, not into trophozoites which are schizonts and produce more merozoites, but into trophozoites which are either male or female. These individuals produce either male or female sex cells. The male and female sex cells of any animal are called *gametes*, and the cells of any animal which produce gametes are called *gametocytes* (fig. 26). The cells of *E. caviae* which grow up to produce its male and female sex cells are, therefore, its gametocytes.

The male gametocyte produces, by a process of division which need not be described in detail here, a large number

---

PLATE III

Fig. 25. Merozoites of *Eimeria tenella*, the cause of caecal coccidiosis of poultry. Size 16 × 2 microns

Fig. 26. Zygotes of *Eimeria perforans* forming oocysts. Size 14·5 × 12·3 microns

Fig. 27. Oocysts of *Eimeria caviae* each containing 4 spores. Size 19 × 16·5 microns

Fig. 28. Oocysts of *Eimeria tenella* each containing 4 spores. Size (average) 22·6 × 19 microns

Fig. 29. *Hypoderma bovis*, the adult female warble-fly. Length 15 mm. Note the ovipositor to the right of the hind limb of the fly

Fig. 30. Lesions on the back of a bull caused by the grubs of the warble-fly

Fig. 31. The mature warble-fly grub in the sac under the skin of a cow

Fig. 32. *Hypoderma bovis*, the mature grub. Length 16–28 mm.

PLATE III

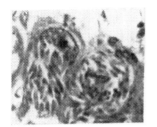

Fig. 25

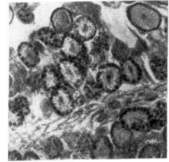

Fig. 26

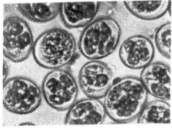

Fig. 27

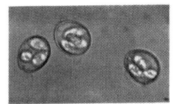

Fig. 28

Fig. 29

Fig. 30

Fig. 31

Fig. 32

of minute, comma-shaped male gametes (sperms) each of which has two small whip-like protoplasmic organs (*flagella*) by means of which it can swim. The female gametocyte does not divide, but stores up food granules in its cytoplasm and becomes, after certain changes have occurred in its nucleus, the female gamete, which corresponds to the unfertilised egg of other animals. The female gamete is then ready to be fertilised.

The male gametes escape from the cells of the host in which they have been formed and find their way to the female gametes. Fertilisation is effected by the entry of one male gamete into the female gamete and the fusion of their nuclei. The female gamete then corresponds to the fertilised egg of other animals and is called the *zygote*.

Immediately after one male gamete has entered the female gamete all the other male gametes are excluded by the formation, out of granules in the contents of the female gamete, of two resistant membranes which completely enclose the fertilised egg. The zygote, thus enclosed in its resistant membranes, is the oocyst with which this description began. The oocyst escapes from the cell of the host in which it has been fertilised and passes out of the host in its droppings. Inside it are formed the sporozoites whose function it is to infect a new host.

This kind of life history differs from that of the roundworms described above in one important respect. During it an alternation of asexual and sexual generations occurs. We shall find that a similar alternation of asexual and sexual generations occurs during the life histories of some parasitic animals which require intermediate hosts. It occurs, for instance, during the life history of the malarial parasites of man, which is very similar to that of *E. caviae*.

One important difference between these two life histories
is the fact that the asexual and sexual generations of the
malarial parasites occur in different hosts, the asexual
generations occurring in the red blood cells of man and the
sexual generations in the mosquito which transmits malaria
to him. The mosquito is therefore the definitive host of
these parasitic animals and man is their intermediate host.

Type I D (figs. 33, 73): *Trichinella spiralis*, the pork trichina-
worm, parasitic in man, the pig and some other animals

We have now selected, from among parasitic animals
which are directly transmitted from one host to another,
examples which show how those phases of the life history
which are not parasitic are protected, during their sojourn
in the world outside the host, by membranes which enclose
them. The pork trichina-worm, however, goes a step
further in parasitism. It secures protection for its larval
phases, not by the formation of resistant membranes around
them, but by the abandonment of the world outside the host
in which they might be exposed to injurious influences. The
same step is taken, we shall find, by species of parasitic
animals whose transmission from host to host is indirect
through an intermediate host, an example of these species
being the human malarial parasites whose life history is
described in Chapter 4.

The pork trichina-worm is a roundworm which is parasitic in a number of mammals, including man.

The adults are small. The adult male measures about
$1\frac{1}{2}$ mm. long and the adult female is about twice as long.
The adult males and females live in the walls of the first
part of the small intestine (*duodenum*) of man, the rat, the
pig, the fox, the bear, the badger and other flesh-eating

mammals. They cause a disease called *trichiniasis* or *trichinosis*, which may be so mild that it is not recognised as a disease at all, or so severe that it may cause prolonged illness or death. The adult worms damage the small intestine of the host, causing nausea, vomiting, diarrhoea

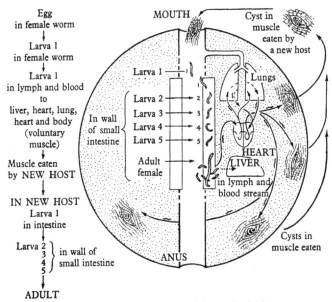

Fig. 33. Diagram of the life history of the pork trichina-worm, *Trichinella spiralis*, Type I D

and abdominal pain. These symptoms are, however, usually less severe than the pain and other symptoms which follow later when the larvae of the trichina-worm leave the intestine and migrate into the muscles.

The female trichina-worm produces eggs, but she does not lay them. Instead the eggs hatch inside her womb (*uterus*), so that she actually lays, not eggs, but living larvae.

Animals which do this are said to be *ovo-viviparous*, in contrast to oviparous animals, on the one hand, which lay eggs, and viviparous animals on the other, which produce living larvae without the formation of an egg capsule.

The larvae of the trichina-worm are set free into the spaces which contain the lymph in the walls of the host's small intestine. From there they reach the lymphatic vessels which are collecting from the intestine the products of the digestion of the fats of the food. These vessels are called lacteals, and they lead to a single larger tube, called the *thoracic duct*, which empties its fatty contents into the large vein collecting the blood from the left side of the head and neck and the left arm (*superior vena cava*). The larvae of the pork trichina-worm thus reach the venous blood stream and are carried by this to the right side of the heart. Some larvae may reach the thoracic duct from the peritoneal cavity in the abdomen, into which they migrate from the intestinal wall. Others may enter the venous blood in the branches of the hepatic portal vein in the intestinal wall, and in this blood they are carried to the liver, from which they reach the right side of the heart.

Whatever the route by which these larvae reach the heart, they leave it again in the venous blood which is going to the lungs and, passing through these organs, they enter the oxygenated blood which takes them back to the heart; but this time they are taken to its left side, from which the arterial blood is pumped all over the host's body. In this arterial blood they are distributed all over the body of the host. Many of them thus reach tissues in which they cannot live, and in these they perish. Most of the ones which survive settle down in muscles which are under the control of the will (voluntary muscles), and especially they settle down in the muscles which move the jaws and those which perform the movements of swallowing and breathing. They are found, therefore, in the muscles which close the jaws (*masseter muscles*), in the midriff (*diaphragm*), which is the transverse sheet of muscle which separates the chest from the abdomen, and in the muscles between the ribs (*intercostal muscles*). They may, however, be found also in other muscles. Often they are found in the muscle which forms a cap over the point of the shoulder (*deltoid muscle*), in the muscles which form the calf of the leg (*gastrocnemius muscles*) and in the muscles which move the eye.

In these muscles the larvae grow and a reaction of the host forms around them a covering or *capsule*, so that they lie in lemon-shaped cysts about 0·4 mm. long and about 0·25 mm. in diameter. Inside this capsule the larva lies coiled in a spiral (fig. 73). Because it is shut off inside this capsule, the larva cannot leave the situation in which it has been encapsuled. It is, therefore, no longer able to influence its own fate. It must wait passively until it is transferred to another host. Like the infective larva of the sheep stomachworm, it can enter another host only through its mouth. Because it cannot by its own efforts leave the muscle in which it is imprisoned, it must wait until that muscle is eaten by the new host.

If, for example, the larvae are lying in the muscles of a rat and that rat is eaten by a wild boar or by a fox or a bear, the larvae will be set free from the muscles in which they are encapsuled by the digestion of these muscles in the stomach of the new host. They are then able to complete their development in the duodenum of the new host. They find their way into the walls of the duodenum and grow up there into the adult roundworms, which produce a new generation of larvae. This second generation of larvae eventually passes to the muscles of the new host and there await the eating of this new host by yet another one.

This kind of life history therefore depends entirely upon parasitic life. The parasitic animal never enters the world outside its hosts and could not survive in it. Exclusion of the external world has the advantage that the parasitic animal is completely protected inside its hosts throughout its life history; but it also involves complete dependence upon the host and its habits, not only for the means of livelihood of the parasitic animal, but also for its transference from host to host.

If, for instance, the larvae of the trichina-worm enter the muscles of the pig, they cannot get from this pig into another host unless the flesh of the pig is eaten uncooked by that host. Man, when he eats pork which has not been cooked sufficiently to kill the larvae, may be infected by them, but, when this happens, larvae which then develop in man and get into his muscles, cannot, under modern conditions, pass from man to any other kind of host, because human flesh is not nowadays eaten either by man or by other animals. The larvae therefore eventually die inside the capsules in which they are enclosed and frequently they become calcified (see Chapter 7).

The number of people in whom they may be found is, in some countries, astonishing. Examination of the midriffs and other muscles of people who have died from various diseases in United States hospitals has, for instance, shown that as many as 16% of these people have been infected, at some time during their lives, with the pork trichina-worm, and in some parts of the United States as many as 27·6 %. Although outbreaks of trichiniasis are rarely reported in England, a recent survey showed that an average of 10·8 % of people who had died from various causes in six hospitals in England had larvae of the pork trichina-worm in their muscles. Even in the Arctic trichiniasis is a serious problem. The source of human infection there may be walrus meat, although the trichina-worm has not yet been found in the walrus. It has, however, been found in 90 % of Arctic sledge dogs, which it may incapacitate, and in Polar bears, Arctic foxes and in one specimen of the bearded seal. Trichiniasis is difficult to diagnose, especially when the number of larvae taken in is small. Frequently it produces symptoms which resemble very closely those of influenza, typhoid fever and other diseases. The measures which are taken to protect people from infection with the larvae of the pork trichina-worm are described in Chapter 10, where the disease which these larvae cause is further discussed.

Under more primitive conditions, of course, man must have infected himself much more frequently than he nowa-days does with the pork trichina-worm, because he cooked his meat less efficiently, probably more frequently ate it raw and also more often ate the flesh of animals other than the pig, such as the bear, from which he infected himself. In former times, when cannibalism was practised, it was possible for man to infect himself by eating other men whose muscles contained living larvae of the pork trichina-worm.

The life histories so far described have been selected to indicate the gradually increasing dependence upon para-sitic life of parasitic animals which pass directly from host to host. Before we go on to study examples of a similar increasing dependence upon parasitic life among species where transmission from host to host is indirect, it is important to consider briefly some variations of the direct types of life history which have just been outlined. These variations are important, not only from the biological point of view, but also because they are shown by parasitic animals which either cause severe disease of man or inflict severe losses upon his civilisation. It will be sufficient for our purpose if we consider briefly here four of these variations. Later on in this book it will be necessary to refer to them again more than once.

Type IIA (figs. 34, 56): *Ancylostoma duodenale*, the human hookworm

The plan of the life history of this species is the same as that of *Haemonchus contortus*, but the behaviour of its non-parasitic infective larvae introduces variations of this plan which have great economic and biological importance.

The non-parasitic infective larvae of the human hookworms are exactly comparable to those of the stomachworm of sheep. The first larva hatches from the egg and grows and moults its skin to become a second larva, which also grows and moults its skin to become a third, infective, larva, which retains the moulted skin of the second larva as a loose sheath around its body. These infective larvae can, moreover, infect man by entering his mouth with his food and drink. They have, however, developed a new method of infection of new hosts. They have developed the ability to bore their way through the host's skin.

When they bore through the skin, the infective larvae of the human hookworm arrive in the tissues under the skin

and there enter the lymph and blood. By the venous blood
stream they are carried first to the heart and then to the
lungs, where the venous blood goes to be aerated. Some of
the larvae pass through the fine capillary blood vessels of
the lungs and pass on with the blood to the various organs

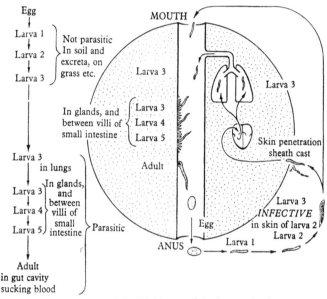

Fig. 34. Diagram of the life history of the human hookworm,
Ancylostoma duodenale, Type II A

of the body. Those which do this fail to develop further and
die. The majority of the larvae, however, remain in the
small capillary blood vessels of the lung. They then break
their way out of the capillaries of the lung and travel up the
air tubes to the *trachea* and thus reach the back of the throat.
There they pass over into the gullet (*oesophagus*), down
which they are carried to the stomach. Passing through the

stomach, they reach the first part of the small intestine. The third larva attaches itself to the bowel wall, moults its skin to become the fourth larva, which grows and undergoes the fourth (final) moult to become the fifth larva. This becomes the sexually mature adult without further moulting. The adult female is 10–18 mm. long by about 0·6 mm. in diameter and the adult male 8–11 mm. by about 0·5 mm. in diameter. Both are most numerous in the middle third of the small intestine (lower part of the duodenum, ileum and jejunum). Usually a human hookworm will be laying eggs some 5 weeks after its infective larva has penetrated the human skin.

The ability developed by this species to enter the host through the skin offers greater opportunities to infect new hosts. It is an adaptation which the larval stages of many flukes and insects have also made. It is considered further in Chapter 9. It carries with it, however, the hazards of the complex journey from the skin to the small intestine which the larvae must perform.

Type IIB (figs. 3, 35, 62): *Ascaris lumbricoides,* the large roundworm of man, the pig and other animals.

The infective larvae of this species of roundworm show another type of migration. They develop inside the eggs, which are about 60 microns long by 42 microns broad.

When these infective eggs are swallowed by a new host, the infective larvae in them, which are about 0·3 mm. long, do not hatch out of them until the egg reaches the first part of the duodenum. Because the adult worms also live in this situation, we might expect that these larvae would grow up here to the adult stage. This, however, they do not do. First they make a complicated journey through the host's body which is similar to the journey made by the larvae of the hookworms. The difference is that the infective larvae of *Ascaris lumbricoides* start their journey from the duodenum.

They penetrate the walls of this organ and then enter the small blood vessels in the walls of the duodenum. The blood in these vessels is collecting the products of the digestion of the food, and all this blood must pass, by way of the hepatic portal vein, to the liver. Infective larvae of *A. lumbricoides* which enter these blood vessels are therefore also carried to the liver. A few of them no doubt bore their way directly from the duodenum through the abdominal organs to the liver, whose outermost protective layer (capsule) they penetrate; but most of them reach the liver by way of the hepatic portal vein.

In the liver they provoke a reaction of the liver tissues against them, and the result of this reaction is that the liver tissue may be altered around the larvae and some of the larvae may be killed by it or enclosed in coverings (capsules), which the tissues of the host make and inside which the larvae may die. Much of the tissue formed as the result of the reaction against these larvae is of the kind called connective tissue, which gives to organs in which it is abundant a whitish appearance. Livers which have reacted vigorously to the larvae of *A. lumbricoides* may therefore show white spots upon their surfaces and in the liver substance and the name *white-spot liver* has been given to this condition.

Most of the larvae of *A. lumbricoides*, however, pass through the liver to the right side of the heart and, from this, they pass with the venous blood to the lungs, where they grow and undergo two moults of the skin. The second larva which hatches from the egg thus changes in the lungs of the host into the third and then the fourth larva. The fourth larva then breaks out of the lung capillaries into the air sacs and travels from these into the air tubes and

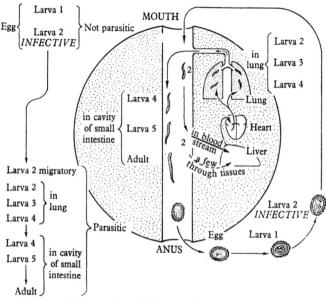

Fig. 35. Diagram of the life history of the large roundworm, *Ascaris lumbricoides*, Type II B

up these to the windpipe. Reaching the junction of this with the gullet in the throat, it passes, as the larvae of the hookworms do, down the gullet and through the stomach to the small intestine from which larval migration began. In the duodenum the fourth larva undergoes the fourth (final) moult to become the fifth larva, which becomes the sexually mature adult without further moults. The time required for the whole development from the emergence of the infective larva to the attainment of the adult state is, in a favourable host, about 4–5 weeks. The adult worms are large and stout, the female being about 20–40 cm. and the male 15–30 cm. long.

The main differences between this life history and that of the hookworms are, therefore:

(1) the development of the infective larva inside the egg, so that an infective egg is produced instead of free-living infective larvae;

(2) the penetration of the wall of the small intestine by the infective larva instead of penetration of the skin.

Type IIC (figs. 29–32, 36, 83, 84): *Hypoderma bovis* and *H. lineatum,* the ox warble-flies.

These two species cause the well-known warbles (fig. 30) of cattle in Europe, including Britain, and in the United States. Their life histories are essentially the same and, being insects, their anatomy and biology are totally different from those of the Protozoa and roundworms so far described. The plan of their life histories, moreover, differs from that of the species just described in one important particular. If we except the pork trichina-worm, the older larval phases and the adults of all the species so far described are parasitic, the eggs and the younger larval phases being non-parasitic. The warble-flies reverse this plan. Their younger larval phases are parasitic, while their older larval phases and the adults are not.

The adult warble-flies belong to the order of insects which is called the Diptera because its members possess only two wings. They belong to the family Oestridae, other members of which are the sheep nasal-flies and the horse bot-flies (see

Chapter 5). The warble-flies are more nearly related to the stable and tsetse flies than to the blood-sucking horse-flies and clegs with which they are sometimes confused.

Hypoderma lineatum and *H. bovis* are both fairly large, dark, hairy flies with bands of yellowish or orange hairs on the abdomen which give them some resemblance to bees. *H. bovis*, which is about 15 mm. long, is rather bigger than *H. lineatum*. Neither species can feed, because the mouth-parts have not developed; but the females of both species have an ovipositor for laying eggs, which consists of the hinder segments of the abdomen telescoped into each other. There is, however, no sting. The warble-fly can neither bite nor sting the cattle. Nevertheless, cattle may be violently upset by the approach of these flies, and they seem to try to avoid them (cf. Chapter 10). Often they seem to listen for the peculiar humming noise made by these flies, which is not loud but can be heard by human ears.

The flies approach the cattle in order to fix their eggs to the hairs. Each of the stalked, elongate-oval eggs has a clasp which fits the hair to which the egg is attached with a cement-like substance (see Chapter 5). The eggs of both species are about 1 mm. long. *H. lineatum* fixes her eggs in rows of about five to twelve eggs in a row, and as many as twenty eggs may be fixed to a single hair (fig. 84). The eggs of this species are laid in batches of 50 to 60 up to a total of 500–800. *H. bovis*, on the other hand, fixes only one egg to each hair (fig. 83).

The egg-laying behaviour of the two species differs in important respects. *H. lineatum* is the stealthier of the two. It usually approaches on the wing, alighting on the ground near a cow and then trying to get near the hind legs by a series of short flights which are almost like jumps; or the fly will alight at once on the cow's heel. If the cow moves away, the fly follows. A cow may be seen to be kicking at the flies and may even knock a fly to the ground; but the fly soon comes back again.

In contrast to this, *H. bovis* makes what have been called vicious and persistent attacks. They are, of course, not attacks in the ordinary sense of this word, because the flies cannot hurt the cattle. All that the flies do to the cattle is to lay their eggs upon their hairs. *H. bovis* seldom alights upon the ground. It approaches the cattle at about the height of the back and 'strikes', as the farmers put it, several times in rapid succession, cementing an egg to a hair at each strike. Then it leaves the cattle for a while before repeating the process.

All observers are agreed that cattle are much more alarmed by *H. bovis* than by *H. lineatum*. The latter species lays eggs especially on the short hairs overhanging the back of the hoof, and because it prefers the heel region and strikes at it, this fly is called by the Americans the 'heel-fly'. It also lays, however, higher up the hind legs around the hock and on the belly, flanks and even the fore legs. When cattle are quiet, eggs are laid also on the escutcheon, base of the udder and tail. *H. bovis* also prefers to lay eggs on the legs, but because it alarms the cattle more and therefore often has to chase them, its eggs are more often laid higher up the hind legs or on the rump.

Egg-laying begins when the adult females have paired. *H. lineatum* may begin to lay late in May, but usually it appears from early June to mid-July. *H. bovis* appears rather later, beginning in late June and persisting later than *H. lineatum*. Both species prefer for their egg-laying quiet, sunny weather, and both avoid shade and refuse to follow cattle across water. For this reason the provision of shaded areas and access to shallow water in which the cattle may cover their feet or lower parts of their legs will help to reduce the number of eggs laid. In windy weather cattle will seek breezy places as if they were aware that in these they will avoid the warble-flies.

Neither shade nor water nor wind will, however, enable the cattle to feed as they should do while the warble-flies are about, and this loss of normal feed may have serious effects upon them. They are restless, and may make sudden rushes to avoid the flies or may stampede and injure themselves as they do so. They have been known to stampede over cliffs or to become bogged in their efforts to get the lower parts of the legs covered in water; and there are records of cattle brought to complete exhaustion by their efforts to avoid the flies. Obviously cattle cannot thrive when they are thus disturbed. Not only do they fail to fatten, but the milk yield may fall. In the United States reduction of the milk yield by 10–25% has been recorded.

When the eggs have been laid, the adult flies cause no further trouble. They live only for about 5–6 days. But from the egg there hatches, in about 3–6 days, a small grub, about 0·5 mm. long, which crawls down the hair and burrows into the skin of the cow. It takes this grub about an hour and a half to burrow right through the skin, and as it disappears beneath it some lymph exudes from the hole that it has made.

Other grubs may enter by the same hole or may make other holes near by, so that a good deal of irritation may be set up in the areas where the eggs have been laid. Small pimples may be formed, and there may be inflammation of the skin with matting of the hair and the formation of scabs.

The cattle react to this condition by vigorously licking the irritated parts and by stamping their feet and other movements. The licking of these irritated areas doubtless gave rise to the idea that the grubs enter the cattle, not through the skin, but through the mouth after they have been licked off the hairs. This idea, which may have been partly due to confusion of the warble-fly with the bot-flies of horses, which do enter their hosts through their mouths, has now been disproved. Experimental work has shown that larvae of warble-flies introduced into the mouths of cattle will not develop into warbles and that cattle do not develop warbles when they are effectively protected from the egg-laying of the adult flies. In addition to this, the entry of the grub through the skin has actually been watched and photographed.

After they have burrowed through the skin the grubs of the warble-flies begin a remarkable sojourn inside the cow's body about which we know very little. It is known, however, that the young grub of *H. lineatum* spends the autumn and early winter wandering about in the tissues of the leg, chest and abdomen of the cow. It then finds its way to the wall of the gullet and can be found here from December until about the middle of February. It is then about 3·13 mm. long or even longer.

From the middle of February for about a month these grubs pass back through the tissues to the side of the spinal column of the back, where the warbles (fig. 30) are found at this time of the year. During this migration backwards some grubs may enter the spinal canal, in which the spinal cord lies. The ones that do this usually come out again without doing any harm, but instances are known of grubs entering the spinal cord itself and causing paralysis of the hind limbs of the cow.

When it reaches the back of the cow the young grub has completed its migrations. Helped possibly by a chemical substance which it secretes, it cuts a hole in the skin of the cow and, placing the two breathing pores (*spiracles*) at its hind end against this hole so that it can

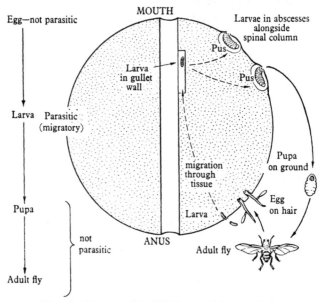

Fig. 36. Diagram of the life history of the warble-fly, *Hypoderma bovis*, Type II C

breathe the air, it settles down to feed and grow. Its presence beneath the skin causes some irritation of the surrounding tissues, which is increased by the spines on the under-surfaces of the grub. The tissues of the cow react to this irritation by forming a fibrous bag (*cyst*) (fig. 31) around the grub, and inside this bag, lying horizontally beneath the skin, the grub feeds upon the exuded tissue fluids of the cow and, if the warble gets infected by bacteria, as it often does, on the matter in the abscess which results.

After about 7–12 weeks of growth in the warble on the back, the grub is mature (fig. 32). The mature grub of *H. lineatum* is 16–26 mm. long, with a diameter of 7–11 mm. and the mature grub of *H. bovis* is rather larger. Some grubs may mature as early as the second half of March, but the majority usually mature throughout April, May and even early June. When they are mature the grubs wriggle out of the warbles and fall to the ground. Here the grub moults its skin, keeping the cast skin as a covering for its pupa or chrysalis stage, inside which the adult fly is formed. By the end of May or at the latest the end of June the last of these grubs to mature will have left the cows. The duration of the pupal stage depends upon the weather and varies between 5 and 8 weeks.

The life history of *H. bovis* is very similar to this, except that the adult flies appear about a month later, being on the wing in July and August. Their larvae therefore appear on the backs of cattle about a month later than those of *H. lineatum* and so are maturing from April until May. It is doubtful whether the young grubs of *H. bovis* pass to the wall of the gullet before they migrate to the back.

H. lineatum and *H. bovis* are usually parasitic only in cattle, and the other species of this genus are also usually restricted to only one species of host. Thus *H. diana* causes warbles of red deer and cannot live in cattle and *H. crossi* is parasitic in Indian goats. Species of the genus *Oedemagena* are parasitic in reindeer.

The host-specificity of *Hypoderma lineatum* and *H. bovis* is not, however, absolute. Both species are sometimes parasitic in horses, and in some parts of England warbles of horses are quite common. It is doubtful, however, whether the maggots of either species ever mature in horses. The maggots of *H. bovis*, *H. lineatum* and *H. diana* have all been found under the skin of man, where they cause lesions similar to the warbles of cattle. The people thus infected are usually associated more or less closely with cattle. The maggots do not normally mature in man, partly because they are removed before they can mature and partly, no doubt, because man is an unusual host in which they cannot mature even if they have the opportunity to do so. Parasites which thus enter unusual hosts in which they cannot survive very long are called *accidental parasites*.

The distinction shown by the life histories of the warble-flies between parasitic young larval phases and non-parasitic older larval phases and adults is characteristic of the family Oestridae, to which the warble-flies, horse bot-flies and the sheep nasal fly belong and also of the flies whose maggots cause myiasis of man and animals (see Chapter 9). It is also characteristic of most of the Gordian roundworms (cf. Chapter 6), the adults of which are not parasitic. They live in fresh water, their larvae being parasitic, first in the aquatic larvae of two-winged insects and later in adult insects and fish. Members of the roundworm family Mermithidae also have this kind of life history and, because these roundworms are biologically so different from insects, one species of them will be briefly described.

This species is well known to gardeners in and around London, Cambridge and in some other parts of Britain. Its name is *Mermis nigrescens*. It has been called the rainworm, because the adults often appear above the soil of gardens and other cultivated land after thunderstorms have wetted the soil in warm midsummer weather, and some people have therefore thought that these worms came down with the rain or were in some other way associated with it.

It is true that they are associated with moisture, as most roundworms are, the reason being that, although most roundworms have considerable power to resist drying, they move about most easily on damp or wet surfaces. For this reason, no doubt, the female rainworms are often seen on the surface of the soil in the early morning before the sun has dried up the dew or rain of the night.

The adult female rainworm is a long, thin, brownish-black worm which may be 12 cm. long. It climbs plants in order to lay eggs upon them, and may climb as high as 3 feet from the ground. After summer rain it may be seen curling and writhing on the surface of the soil, where there may be small masses composed of several inter-twined worms, or the worms may extend themselves into the air from the surface of the soil and gently wave to and fro, looking as if they were in search of a plant to climb.

The eggs are laid upon wet plant surfaces and are left there by the female. When they hatch, they liberate a larva which begins the parasitic phase of its life history by entering the body of an arthropod host, which is very often an insect. It is probable that we do not yet know all the possible hosts of M. nigrescens, but in Britain its larvae often enter the bodies of earwigs. In other countries they are parasitic in the bodies of various species of grasshoppers and in cockchafers.

Inside the body of the insect host the larvae of M. nigrescens grow until they may almost fill the interior of the insect's body. The damage they do to the insect host may kill it, so that M. nigrescens helps to control the damage done to cultivated plants by earwigs. When its larva is fully grown, it leaves its insect host and enters upon its non-parasitic life in the soil.

The larvae of other species of the genus *Mermis* are parasitic in various insects belonging to the orders Orthoptera and Lepidoptera.

One feature of this kind of life history is the long period of time which it requires. As long a time as two years may elapse between the emergence of the parasitic larval stage from its insect host and the emergence of the female from the soil to lay eggs. During all this time the males and females live deeply buried in the soil. The males, in fact, seem to leave the soil only rarely. They can be found only by patient digging and patient search of the soil dug up.

CHAPTER 4

REPRESENTATIVE LIFE HISTORIES, II

In this chapter some representative life histories of species which use one or more intermediate hosts will be described.

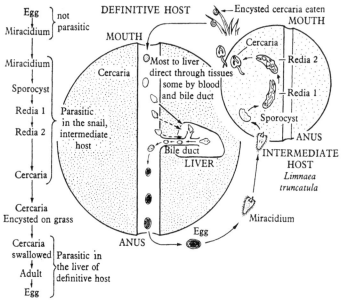

Fig. 37. Diagram of the life history of the liver fluke of cattle and sheep, *Fasciola hepatica*, Type III A

When a parasitic animal has committed itself to the necessity of undergoing a part of its development inside the body of an intermediate host and the rest of its development inside the body of a definitive host, it increases the risk that it may not be able to complete its life history, but it gains certain advantages.

The intermediate host is usually closely associated with the definitive host, and this association helps the parasitic animal to infect more definitive hosts and also to spread its

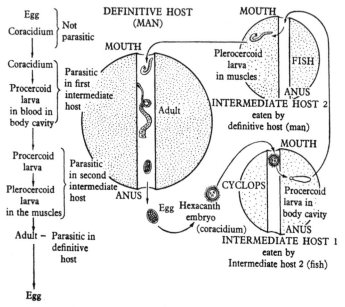

Fig. 38. Diagram of the life history of the human broad tapeworm, *Diphyllobothrium latum*, Type III B

species over a wider area. When one of these two kinds of host is an insect, the extension of the geographical distribution of the parasitic animal effected in this way may be considerable.

The intermediate host protects and nourishes the delicate larval phases of the parasitic animal and the multiplication, mentioned in the previous chapter, of the number of individuals derived from each fertilised egg of the parasitic animal, occurs in it. The reader is reminded of this multiplication here, because it has, as later chapters will show, an important bearing upon the study of diseases caused by parasitic animals. It is illustrated by the life histories which follow.

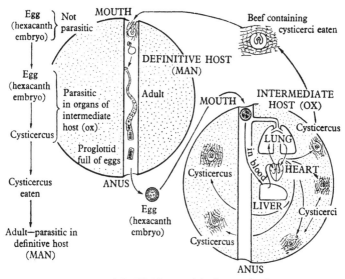

Fig. 39. Diagram of the life history of the human beef tapeworm,
Taenia saginata, Type III C

Type III A (figs. 1, 37, 40–44 and 83): *Fasciola hepatica*, the liver fluke of sheep, cattle and other animals.

The definitive hosts of this species may be sheep, goats, cattle and other animals that chew the cud, pigs, hares, rabbits, beavers, elephants, horses, dogs, cats, kangaroos, and rarely man. Its usual definitive hosts are, however, sheep, cattle and other ruminant animals. The adult liver fluke has a flattened oval body tapering to a point at each end, so that it is shaped rather like a leaf. It is about

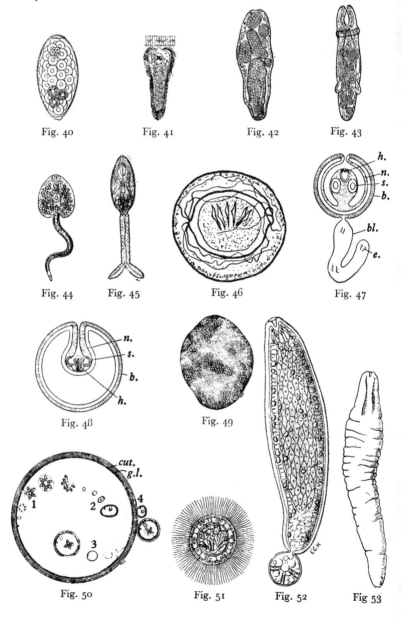

Fig. 40   Fig. 41   Fig. 42   Fig. 43

Fig. 44   Fig. 45   Fig. 46   Fig. 47

Fig. 48   Fig. 49

Fig. 50   Fig. 51   Fig. 52   Fig 53

30 mm. long and about 13 mm. broad at its broadest point. It is parasitic, not in the food canal of its hosts, but in the canals (bile ducts) which conduct the bile out of the liver.

Its eggs (fig. 40) are ovoidal and are about twice as large as those of the sheep stomachworm; they measure about 140 microns long and are about half as broad. They are provided, as the eggs of many other species of flukes are, with a lid (*operculum*) at one end and are passed out of the liver with the bile into the duodenum and reach the outer world in the droppings of the host.

When the conditions in the outer world are favourable, the lid opens and a minute larva, called the *miracidium* (fig. 41) larva, hatches out. It is a pear-shaped organism,

---

FIGS. 40–53

Figs. 40–44. The egg and larvae of the liver fluke of sheep and cattle, *Fasciola hepatica*. Fig. 40. Egg, showing its lid (*operculum*). Size 140 × 76 microns. Fig. 41. Miracidium, showing the cilia covering it and the papilla which helps it to penetrate into the snail. Size 130 × 27 microns. Fig. 42. Sporocyst, showing rediae developing in it. Length 0·5–0·7 mm. Fig. 43. Redia, showing cercariae developing in it. Length 1·3–1·6 mm. Fig. 44. Cercaria. Size of the body 0·3 × 0·2 mm. with a tail 0·6 mm. long

Fig. 45. Cercaria of the blood fluke of cattle, *Schistosoma bovis*. Note the forked tail. Length, including the tail, 0·4 mm.

Figs. 46–53. Larval forms of tapeworms. Fig. 46. Egg of *Hymenolepis nana* showing the embryo with its six hooklets. Diameter 30–47 microns. Fig. 47. Cysticercoid larva (diagrammatic): *b*. bladder; *bl*. tail, formed from the wall of the embryo, whose hooklets (*e*.) are seen inside it; *h*. head; *s*. sucker, and *n*. neck of the adult tapeworm. Fig. 48. Cysticercus larva (diagrammatic): *b*. bladder, *h*. hooks, *s*. sucker and *n*. neck of the future adult tapeworm. Fig. 49. Coenurus type of bladderworm of *Taenia multiceps* with many tapeworm heads. Diameter 5 cm. or more. Fig. 50. Hydatid type of bladderworm with daughter cysts (diagrammatic): *cut*. cuticle; *gl*. germinal layer; 1, development of brood capsules and scolices; 2, transformation of scolex into endogenous daughter cyst; 3, formation of endogenous daughter cyst from germinal layer; 4, formation of exogenous daughter cyst. Size variable (see Chapter 8). Fig. 51. Coracidium larva of *Diphyllobothrium latum*. Fig. 52. Procercoid larva of *Diphyllobothrium latum*. Length 0·5 mm. Fig. 53. Plerocercoid larva of *Diphyllobothrium latum*. Length up to 6 mm.

about 130 × 27 microns long, which has two eye spots and swimming hairs (*cilia*), by means of which it swims blunt end foremost in water into which it is hatched. The miracidium must enter the body of the intermediate host, and if it does not do this within 8–24 hours after it has been hatched, it dies.

In Britain the usual intermediate host of this species of liver fluke is the small, amphibious, air-breathing snail called *Limnaea truncatula*. It has been shown that *L. stagnalis* can also act as the intermediate host and possibly *L. pereger* also can. In the laboratory the development of the larvae of *Fasciola hepatica* may occur in *Limnaea palustris* and *L. glabra*; but there is little doubt that the usual intermediate host is *L. truncatula*.

Thus in Britain *Fasciola hepatica* is practically restricted to one species only of intermediate host, but in other countries it has become adapted to other species of snails belonging to more than one genus (cf. Chapter 7).

The miracidium bores its way into the snail by spinning on its long axis and driving in the papilla on its anterior end. It may enter at any point on the snail's surface, but often it enters near the bases of the tentacles or near the opening through which the snail takes in air. It enters the snail's lymph channels and blood vessels and migrates by way of these to some part of the snail's body in which food is abundant. Usually it goes to the digestive gland at the summit of the spire of the shell and here it changes into the next phase of the life history.

This second phase is a sac packed with cells which can give origin to germ cells and it is called the *sporocyst* (fig. 42). When it is full-grown, it is 0·5–0·7 mm. long. From the germ cells in it a third kind of larva, called a *redia* (fig. 43) is formed. The redia is a small, cylindrical organism about 1·3–1·6 mm. long when it is full-grown, which has a mouth and a simple intestine. Many rediae arise inside each sporocyst, and they eventually rupture the walls of the sporocyst and are set free in the tissues of the snail, through which they wander, feeding on the snail's tissues and causing damage which may kill the snail. The rediae, like the sporocysts, contain germ cells, and from these, under certain adverse circumstances, more rediae may be formed.

If the snail survives, the germ cells of the rediae produce, not more rediae, but the next phase of the life history, which is a fourth and final type of larva called a *cercaria* (fig. 44). The cercaria is a discoidal organism about 0·3 mm. long and 0·2 mm. broad, with a tail about 0·6 mm. long. It is therefore just visible to the naked eye and looks rather like a minute tadpole. It already possesses some of the organs of the adult liver fluke. It leaves the redia through an opening called the birth pore, migrates through the tissues of the snail and out into the water in which the snail lives. In this water it swims by means of its tail for an hour or two and then it climbs up grass (fig. 82) or other vegetation. Here it sheds its tail and forms a protective covering (cyst) around itself.

These cysts have a diameter of about 0·2 mm. Some cercariae encyst on parts of plants that are under water and a few encyst on the surface of the water. Experiments have shown that one thousand cercariae can find room to encyst on a single blade of grass. Inside its cyst the cercaria undergoes structural changes. When these are complete it is called a *metacercaria*. It is now ready to infect a new host. The metacercaria must wait in its cyst until it is eaten by a sheep or other definitive host.

The metacercaria is the only phase of the life history which can infect a definitive host, and the metacercaria of this species of liver fluke can infect it only through its mouth. The cercariae of some species of flukes, such as those of the blood flukes (see Chapter 9) which infect man, can, however, infect their definitive hosts by penetration of the skin as well as through the mouth, just as the infective larvae of the hookworms can. Although the metacercaria of the sheep liver fluke becomes passive inside its cyst, it has, before it ceases to be active, placed itself by its own efforts in a position which is favourable to its chances of being swallowed by the definitive host.

Inside their cysts metacercariae can survive, if conditions are favourable to them, for 12 months. They can therefore survive throughout the winter, if this is not too severe, so that they can infect sheep and other hosts which graze on the pastures during the spring following the autumn during which the metacercariae were formed. The meta-cercariae cannot, however, resist drying for long, and they are relatively quickly killed by sunshine. For this reason they do not live for longer than a few weeks on dry hay; but on damp hay they can live for some months, so that animals can be infected with damp hay, even if they never graze upon pastures on which metacercariae have developed.

When it is swallowed by the right kind of definitive host, namely, a sheep or one of a number of other animals, in-cluding man, in which the adult *Fasciola hepatica* can live, the cyst surrounding the metacercaria is digested off by the digestive juices of the definitive host and the metacercariae bore their way through the wall of the definitive host's intestine. They enter the space between the food canal and the other abdominal organs (*body cavity*) and, after about 3 days, they reach the liver and bore through its covering (capsule). They actively bore through the tissues of the definitive host and are not carried, as the infective larvae of the large roundworm of the pig, *Ascaris lumbricoides*, are, by the blood stream to the liver. It is probable, however, that some metacercariae do reach the liver by way of the blood stream. Experiments have shown that the metacercariae may reach the liver 48 hours after they have arrived in the host's intestine.

Once in the liver the metacercariae migrate through its substance for a month or longer, causing bleeding and other forms of damage, and eventually they reach the bile ducts. In the bile ducts they grow and become sexually mature some 10–12 weeks after the metacercariae have been swallowed. The development in the snail requires 6–7 weeks, so that the whole life history, from the egg of one generation to that of the next, requires about 4–5 months. If the climatic conditions are favourable, it may be completed in 2–3 months.

The usual sequence of events in Britain is that the snails begin to become active from March onwards and are infected during the spring. The miracidial larvae which infect them may come from eggs voided that year by sheep or cattle infected with the fluke during the previous summer or autumn; or from eggs passed out during the previous year which have survived the winter's climatic effects. If these climatic effects have not been severe, and if the spring weather is warm and damp and otherwise favours hatching, large numbers of miracidia may hatch out of eggs which have thus survived.

During the summer the snails produce two, or, if the
climatic conditions are favourable, three generations of
young and in these large numbers of cercariae may
develop to infect sheep and cattle grazing throughout the
summer. By the autumn, and not usually before then,
enough flukes may have been established in sheep or
cattle to cause disease, so that fluke disease occurs chiefly
during the late autumn and winter. Some evidence of it
may be seen earlier in the year if animals have been put to
graze upon infected pastures from which they have taken
up metacercariae formed during the previous autumn and,
at any time during the summer and autumn, an acute
form of fluke disease may develop suddenly in sheep and
this may kill them in a few days.

Four features of the life history which has just been
described need emphasis here. First, neither the egg, the
miracidium nor the encysted metacercaria is parasitic, so
that the parasitic animal has not yet abandoned the world
outside the host. Secondly, there is an alternation of
sexual and asexual generations. Thirdly, the asexual and
sexual generations are in different hosts. Fourthly, there
is a multiplication of the number of individuals derived
from each fertilised egg comparable to that which occurs
during the life history of *Eimeria caviae*.

Although the particular species of liver fluke just described produces from each fertilised egg only one miracidium, and although from this only one sporocyst arises,
each sporocyst produces several rediae and each redia may
produce more rediae, each of which may give origin to more
than one cercaria. The number of cercariae derived from
each fertilised egg is therefore considerable. Each egg of
some species of flukes may, indeed, produce in this way

many thousands of cercariae. The numbers produced and the biological significance of this process of multiplication of the number of individuals derived from each egg are discussed in Chapter 6.

The life history of *Fasciola hepatica* has been chosen as a type of fluke life history, but it is important to realise that all species of flukes do not need an intermediate host. An important group of them mentioned in Chapter 5 are directly transmitted from host to host.

Among those which do need an intermediate host, some species can use more than one species of animal as their intermediate hosts, and some species need, not one intermediate host, but two successive ones; they must pass, that is to say, from one intermediate host to another, and must pass part of their development in this second intermediate host before they can pass on to the definitive host. Some species of fluke, for example, pass first into a snail and then into either a crustacean animal, the larva of an insect, a fish, or even into the tadpoles of the common frog or the frog's skin, before they can infect their definitive hosts. This necessity of using two independent intermediate hosts introduces us to the next type of life history to be considered.

Type III B (figs. 38, 51–53, 78): *Diphyllobothrium latum*, the fish or broad tapeworm of man.

The plan of the life history of this species of tapeworm is similar to that of the liver fluke, but it is more passive. The tapeworm relies for its entry into the intermediate and definitive hosts, not upon the activity of its larval phases, but upon the fact that these hosts normally eat one another. Some of the larval phases of this particular tapeworm, moreover, are, like some of the larval phases of the liver fluke, not parasitic, although other species of tapeworms have reduced the non-parasitic phases of their life histories so much that all phases of them are parasitic except the eggs.

The adult fish tapeworm lives in the intestine of man and many other species of mammals, such as the dog, cat, fox, bear, pig, walrus, seal and sea lion. Its body, like that of other tapeworms, consists of a chain of flattened pieces which are called segments (*proglottides*) (cf. fig. 2).

The chain grows to different lengths in different species of definitive host, or in different individuals of the same species of definitive host. The length of the fish tapeworm varies from 7 to 60 feet. Its segments are rather broader than they are long and there are 3000 to 4000 of them. They grow, as the segments of all tapeworms do, from the neck of the tapeworm which follows the minute head. The head is at the narrowest end of the tapeworm, and, because the segments grow bigger as they get older, the youngest ones, which are being budded off from the neck, are the smallest. The whole tapeworm therefore is narrowest at its head end. There is, however, in this species, no clear distinction between immature segments nearer the neck and more mature ones lower down the chain, as there is in tapeworms of the genus *Taenia* described below.

The head of all species of tapeworms attaches the whole worm to the tissues of the host and usually it holds on to the host by means of suckers. The heads of many species have hooklets also which help the suckers to maintain a hold (see Chapter 5). The head of the fish tapeworm (fig. 78), however, has no hooklets. It is almond-shaped, and measures only about 2·5 mm. long by about 1 mm. broad. It has two elongated slit-like suckers, called *bothridia*, with which it holds on to the host.

Each segment contains both male and female reproductive organs, so that the fish tapeworm, like all other tapeworms, is hermaphrodite. These reproductive organs mature as the segments grow older, so that the older segments, towards the hinder end of the tapeworm, produce large numbers of sperms and eggs. Most tapeworms produce large numbers of eggs (see Chapter 6) and the fish tapeworm is no exception to this rule. Its egg-production is practically continuous and its egg-laden segments, unlike those of the beef tapeworm described below, are not usually detached from the parent chain while this is in the human intestine. The eggs of the fish tapeworm are ovoidal and each has a lid (*operculum*). They measure about 0·07 mm. long by about 0·05 mm. broad and are discharged from the uterus of each mature segment into the contents of the host's intestine, with which they escape to the exterior. Each egg has a rather thick, golden-yellow, resistant shell. In water at a temperature of 15–25° C. an embryo develops in the egg in about 11–15 days. This embryo is covered with cilia, and has, like the embryos of many other species of tapeworms, six chitinous hooklets inside it. The embryo of a tapeworm which has six hooklets is called a *hexacanth embryo* (cf. fig. 46). The embryo and egg-shell together constitute an *oncosphere*. The free-swimming embryo of the fish tapeworm after it has hatched out of the egg is, however, called a *coracidium* (fig. 51) to distinguish it from the embryos of other species of tapeworms.

The coracidium swims about in the water in which it is hatched out, and unless it meets with a suitable first intermediate host within 12 hours or so, it dies. It does not, like the miracidium of the liver fluke, actively enter its first

intermediate host. It must wait until it is swallowed by one
of the small aquatic animals related to the crabs and lob-
sters which are called *copepods*. The only species of these
which can act as intermediate hosts of this species of tapeworm all belong to two genera only, namely, *Diaptomus* and
Cyclops (fig. 103). In Europe the species of these two genera
which are usually the intermediate hosts of the fish tapeworm are *Diaptomus gracilis* and *Cyclops strenuus*, although
two other species of *Diaptomus* can be used. In North
America species of *Diaptomus* only are used.

When it has been swallowed by these copepods the coracidium bores its way through the intestinal wall into the
space containing blood which surrounds the food canal of
these animals, and there it develops in 2 or 3 weeks into the
next phase of the life history, which is a larva called
a *procercoid larva* (fig. 52), which is about 0·5 mm. long.

No development beyond this stage can occur inside the
copepod. Further development can occur only inside the
body of the second intermediate host, which is a fresh-water
fish. When this fish eats the copepod containing the procercoid larva, the procercoid larva develops, in the
muscles of the fish, into a third larva called the *plerocercoid
larva* (fig. 53) or *Sparganum*, which may reach a length of
6 mm. If the muscles of the fish containing the plerocercoid larva are eaten by another fish, the plerocercoid
larva can become parasitic in the muscles of this second
fish and it may pass in this way to successive carnivorous
fish hosts, until finally the infected fish host is eaten by
a definitive host, in whose food canal the plerocercoid
larva can become the adult fish tapeworm (cf. Chapter 8).

The way in which man infects himself with this tapeworm
will now be evident. The species of fresh-water fish which

can act as the second intermediate hosts of the fish tape-worm include the pike, the perch, the salmon, the trout, the lake trout, the grayling and the eel in Europe, and re-lated species in other countries. If man eats these fish raw or without cooking them sufficiently to kill any plerocercoid larvae which may be present in their muscles, the plerocer-coid larva grows up in the human small intestine into the adult tapeworm.

This is the only way in which man and the other definitive hosts of the fish tapeworm can infect themselves, for the plerocercoid larva is the only phase of the life history which can infect the definitive host and it can infect it only by way of the mouth. Man can therefore easily avoid infection with this tapeworm by cooking all his fish sufficiently to kill the plerocercoid larvae. The control of this infection is thus similar to the control of infection with the trichina-worm.

A few years ago biologists working in Dublin and Liver-pool showed that species of tapeworms which have life histories similar to that of the fish tapeworm live in the intestines of some kinds of birds, such as the cormorant and the herring gull, which feed upon fish, and that the first intermediate hosts of these tapeworms are the same as those of the fish tapeworm, namely, copepods belonging to the genera *Diaptomus* and *Cyclops*. It is important, therefore, to be able to distinguish between the plerocercoid larvae of these species whose adults live in birds and those of the fish tapeworm; but it is by no means easy to do this. The plero-cercoid larvae in the fish must also be distinguished from the larvae of species of roundworms which may be parasitic in the fish.

We have now considered two kinds of parasitic animals which require different kinds of non-parasitic animals as their successive first and second intermediate hosts, namely the flukes mentioned above and the fish tapeworm. The tapeworms next to be described do not require a second intermediate host. They pass directly from the first intermediate host to the definitive host.

Type III C (figs. 39, 72): *Taenia saginata*, the beef tapeworm of man.

This species of tapeworm lives in the small intestine of man. Its length varies from 13 to 40 ft. although specimens as long as 80 ft. have been found. It produces 1000–2000 segments.

Its structure is, in general, like that of the fish tapeworm, but its head is globular and has a diameter of 1·5–2 mm. It has four suckers which are hemispherical, and not elongated. The head, like that of the fish tapeworm, lacks hooklets. The younger segments of the beef tapeworm are, like those of the fish tapeworm, broader than they are long, but, as the hermaphrodite sexual organs in them mature, they lengthen and when they are sexually mature they are longer than they are broad and then measure about 6 mm. broad by 12 mm. long.

These older segments, which are full of fertilised eggs, are often detached singly from the parent chain and may then be discharged from the host in its excreta. They are active after they have left the host and may crawl about on the skin of man or on objects with which the skin comes into contact, discharging, as they move along, a milky fluid which contains the fertilised eggs. In this respect the beef tapeworm differs from the pork and fish tapeworms.

The eggs are smaller than those of the fish tapeworm,

measuring about 0·03 by 0·043 mm. (31 by 43 microns). From them hatches out a six-hooked embryo which is not provided with cilia, so that it cannot swim about. This is one detail which illustrates the much more passive nature of the life history of many tapeworms. Perhaps this is a reason why the embryo does not hatch out of the egg until the egg has been swallowed by the intermediate host. The intermediate host is usually the ox, although the buffalo, the llama and the giraffe can also act as intermediate hosts. The larvae of this species of tapeworm can also be reared experimentally in sheep and young goats.

In the first part of the small intestine (*duodenum*) of the ox the six-hooked embryo hatches out of the egg and penetrates by means of its six hooklets through the intestinal wall. It enters the blood stream, so that embryos of the beef tapeworm may be carried by the blood all over the ox's body. Usually they settle down in the muscles of the ox, where they develop into small bladders, full of fluid. Their dependence on transport about the body of the intermediate host by its blood stream is another detail which indicates the passive nature of the life history.

The wall of each of these small bladders now becomes turned in at one point to form a small projection into its interior, and on this projection a single small tapeworm head is formed. In this way arises the phase of the life history which is called the bladderworm (*cysticercus*) (fig. 48). The bladderworm of the beef tapeworm is oval, greyish-white or opalescent, and about 4–6 mm. long by about 7·5 mm. broad. Because it is found in the muscles of the ox it is called *Cysticercus bovis* (fig. 72). In Britain the numbers of these cysticerci found in the ox may be few and they are found most often in the muscles of the jaws,

heart, midriff, tongue, neck and hind-limb, or in the walls
of the gullet; but they may also be found in other muscles.

When man eats beef infected with these bladderworms
which has not been cooked or otherwise treated to kill the
bladderworms, the bladderworms are digested out of the
beef and the head of the tapeworm inside them turns right
side outwards and attaches itself by means of its suckers to
the wall of the host's small intestine. It then produces the
chain of segments. This species of tapeworm may form one
to two thousand of these.

During this life history the egg is the only phase which
enters the world outside the host; each egg gives origin to
one larva (bladderworm) only, and each bladderworm
gives origin to only one adult tapeworm. This parasitic
animal is, moreover, restricted to one species of definitive
host only, namely, man, and usually uses only one species
of intermediate host also, although it can, on occasion, use
also the other species mentioned above.

Very similar to the life history of the beef tapeworm is the
life history of the pork tapeworm, *Taenia solium*, of which
man is also the only definitive host.

This species resembles the beef tapeworm in appearance
and structure, but its immature segments are broader than
they are long, the egg-laden ones being the reverse—
longer than they are broad. Short chains of these egg-
laden segments are often detached to be passed out of the
host in its excreta; but these are not able to crawl about
outside the host, as the single detached segments of the
beef tapeworm can.

The whole pork tapeworm is shorter than the beef
tapeworm, reaching a length of 6–20 ft., and it does not
form more than 800–1000 segments. The head (fig. 77) is

globular and has, in addition to four suckers like those of the beef tapeworm, about 22–32 small hooklets arranged in a double row, which help the suckers to attach the tapeworm to the small intestine of man.

The usual intermediate host of the pork tapeworm is the pig, but it is an important fact that man, who is its only definitive host, can be its intermediate host as well. Although some kinds of monkeys, and also sheep and dogs, can be additional intermediate hosts, the usual ones are man and the pig.

Man always infects himself with the adult tapeworm by eating the muscles of the pig, in which the bladderworms, to which the name *Cysticercus cellulosae* (fig. 71) is given, have developed after the pig has eaten the eggs passed out in human excreta. The size of these bladderworms varies from 5–8 by 6–20 mm. Their structure is like that of the bladderworms of the beef tapeworm and tapeworm heads capable of infecting a definitive host develop in them 10 weeks after the eggs have been ingested by the pig. They are more numerous in the muscles of the pig than the bladderworms of the beef tapeworm are in those of the ox.

The fact that man himself can be the intermediate host as well as the definitive host of the pork tapeworm requires emphasis. If man should swallow the eggs of the tapeworm living in his own intestine when they get on to his fingers or into his food or drink, or if he should swallow the eggs of pork tapeworms parasitic in other human beings or in pigs, the eggs develop in the human body into bladderworms similar to those of the beef tapeworm and containing only one tapeworm head. These settle in various organs and may cause various symptoms of a disease which is called *cysticercosis*, because it is caused by these cysticerci. Thousands and even tens of thousands of these bladderworms have been found in the human body.

Two of the many other interesting features of tapeworms may be mentioned here. One is the fact that the formation of successive segments of the chain (*strobilisation*) does not usually begin until the tapeworm establishes itself in the definitive host. The strobilisation of one species of tapeworm, however, namely *Taenia taeniaeformis*, which is parasitic in the intestine of the cat and some of its wild relatives, begins in the larval bladderworm, which is found in the livers of rats and mice, which are the intermediate hosts of this species, so that this bladderworm contains, not an inverted tapeworm head only, but a head with a short chain of segments already formed. A bladderworm of this kind, which contains segments already formed, but without sexual organs, is called a *strobilocercus*. The strobilocercus of *Taenia taeniaeformis* is also called *Cysticercus fasciolaris*.

The other feature of tapeworms has considerable practical importance. The fact that each egg of the beef and pork tapeworms, like each roundworm egg, gives rise to only one adult of the species is biologically important, but it is not characteristic of all species of tapeworms. There are some species of tapeworms whose eggs each give origin to many more adult individuals than one, just as those of the liver fluke and many species of flukes do. The significance of this fact is further discussed in Chapter 6; but a brief description will be given here of two different methods by which some species of tapeworms thus multiply the number of adult individuals derived from each of their eggs. These two methods are illustrated by the life histories of *Taenia multiceps*, which causes a disease of sheep and other animals called sturdy, gid, staggers and other names; and that of *Echinococcus granulosus*, which produces the dreaded hydatid cysts which may infect man.

Taenia multiceps (*Multiceps multiceps*) multiplies the number of adult individuals which arises from each of its eggs by forming a number of tapeworm heads inside each of its bladderworms. Lining the bladderworm is a germinal membrane from which these numerous heads arise. A bladderworm which thus produces a number of tapeworm heads is called a *coenurus* (figs. 49, 68–70). When the coenurus cyst is swallowed by the definitive host, each of the numerous tapeworm heads inside it can give rise to an adult tapeworm. As the coenurus of *T. multiceps* may contain several hundred tapeworm heads, the chances of the survival of the adult in the definitive host are considerably increased.

The adult tapeworm lives in the small intestine of the dog. It can also live in the intestine of the wolf. The eggs, passed out in the droppings of the dog, are eaten by sheep, goats, cattle, horses and some other animals which chew the cud, and they can also develop in monkeys. Rarely they develop in man, three cases only of human infection with them being known. These animals are the intermediate hosts, so that this species of tapeworm uses a variety of animals as its intermediate hosts. When the bladderworms have developed in the intermediate host, they must be eaten by the definitive host, the dog or wolf.

The bladderworm of this species normally settles in the brain and spinal cord of the intermediate host, and the name *Coenurus cerebralis* (fig. 68) has therefore been given to it. It can, however, develop in other organs. It grows to a much larger size than do the bladderworms of the beef and pork tapeworms, reaching a diameter of 5 cm. or more. Its pressure upon the brain or spinal cord may cause the symptoms further discussed in Chapter 8. They include inco-ordination of the movements of the intermediate host and loss of the sense of direction, so that the animal moves aimlessly or round and round in circles or staggers about.

Other species which produce coenurus cysts containing many tapeworm heads are:

Taenia serialis (*Multiceps serialis*), the adult of which is parasitic in the intestine of the dog and fox, the coenurus bladderworm (figs. 69, 70) being normally found in the connective tissue of rabbits and hares, though it may also be found in some other rodents and occasionally in man.

Taenia gaigeri (*Multiceps gaigeri*), the adult of which is parasitic in the intestine of the dog, its coenurus bladderworm being found in the connective tissue, brain, spinal cord and other organs of the goat.

The species of tapeworm which illustrates the second method by which tapeworms multiply the number of adult tapeworms derived from each fertilised egg is *Echinococcus granulosus* (*Taenia echinococcus*).

The adult of this species also lives in the intestine of the dog, but it may also infect the fox, cat, and other wild carnivorous mammals. It is a very small tapeworm, consisting of only three or four segments, the whole measuring only 3–9 mm. long. It attaches itself to the fingerlike projections (*villi*) on the wall of the small intestine of the dog, and its life history is similar to that of the beef and pork tapeworms.

Its intermediate hosts are numerous species of mammals. From the human point of view the most important of them is man himself, because the bladderworms of this species are the hydatid cysts (figs. 50, 67) which may produce by their pressure upon various organs the serious effects further discussed in Chapter 8. Other intermediate hosts are sheep, cattle, pigs and, less frequently, horses, camels, monkeys, some kinds of goats, the Asiatic elephant, the tapir, moose, zebu, kangaroo, mongoose, cat, leopard, squirrel, rabbit and dog. This is therefore a species which uses a very wide range of intermediate hosts.

The method used by this species to multiply the number of individuals which arise from each of its eggs is more efficient than that used by *Taenia multiceps*. The germinal layer lining its bladderworm (hydatid cyst) does not direct-ly produce tapeworm heads, as that of the coenurus does. It produces instead numerous stalked, bladder-like struc-tures called *brood capsules* (fig. 50), and numerous tape-worm heads are produced on the walls of each of these (see Chapter 6).

The life history of one other species of tapeworm will be briefly mentioned here, because it is the only tapeworm life history known during which an intermediate host does not occur. This species is the dwarf tapeworm, *Hymenolepis nana*, which is parasitic in man and especially in human children (see also Chapter 8). A species given the same name is parasitic in mice, but the identity of the two is disputed. *H. nana* of man is a small species, which is only about 25–40 mm. long. It lives in the small intestine of man. If its eggs are swallowed, the embryos in them hatch out and enter the walls of the duodenum, in which they develop into a kind of bladderworm, which later becomes the adult tapeworm in the same host. The transmission is therefore direct, like that of the sheep stomachworm, and in this respect this species is unique among tapeworms.

Up to this point we have been considering in this chapter the life histories of species which, with the exception of the dwarf tapeworm, still send some of their larval phases into the world outside the host. The liver fluke and the fish tapeworm are examples of species which enter this outside world twice during their larval development—once in order to reach the intermediate host and again in order to reach the definitive host. The other species of tapeworms whose life histories have been described enter the outside world only in order to reach the intermediate host, the second period of life in the outer world having been eliminated. Let us now consider a type of life history during which the parasitic animal enters the outside world only for a comparatively very brief period just before it enters the body of the definitive host. After that we can consider a life history during which none of the phases of the parasitic animal enters the outer world at all.

Type IV A (fig. 54): *Wuchereria bancrofti*, Bancroft's filarial worm.

This is a species of roundworm which causes a serious disease of man called filariasis, one result of which may be the thickening, enlargement and swelling of certain tissues which is called elephantiasis. Its scientific name, *Wuchereria bancrofti*, commemorates the devoted work done upon this disease by two men, Wucherer and Bancroft. Man is the only definitive host of this species of roundworm; but its intermediate hosts are various species of blood-sucking mosquitoes which belong to more than one genus.

The adult roundworms look like creamy white threads

and they belong to the group of roundworms called the Filarioidea, a name which refers to the fact that all these worms are thin and threadlike. The male Bancroft's filarial worm is about 40 mm. long and about 0·1 mm. in

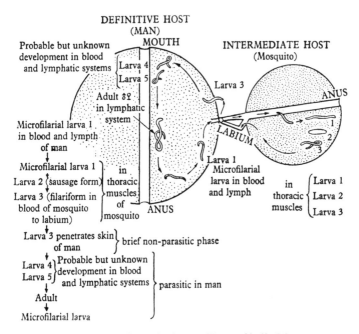

Fig. 54. Diagram of the life history of Bancroft's filarial worm, *Wuchereria bancrofti*, Type IV A

diameter. The female may reach a length of 100 mm. with a diameter of about 0·3 mm. The female is thus, like the females of all roundworms, larger than the male.

The males and females are parasitic especially in the glands and vessels of the lymphatic system of man, and especially in those of the reproductive organs, the region of

the hips (pelvic region) and of the arms and legs. The
female worm produces eggs which have very thin shells
and these are fertilised inside the female. The larva which
develops inside the very thin egg-shell does not hatch out
of the shell, but keeps the thin egg-shell around it, so that
it is enclosed in a very thin sheath. The young sheathed
larva thus produced is called the *microfilaria*. It is about
127–320 microns long by 7–10 microns in diameter.

The microfilariae of Bancroft's filarial worm are discharged from the female while the latter is inside the
human body, and they migrate into the lymph and blood
vessels and circulate about the body of man in the blood
and lymph. They appear in what is called the peripheral
circulation of the blood—in the blood, that is to say,
circulating in the blood vessels of the skin and outer parts
of the human body. In this blood they appear, moreover,
only during the night, and the microfilariae are especially
numerous between the hours of 10 p.m. and 2 a.m. If,
however, the human host sleeps by day and is active either
at work or in other ways at night, the microfilarial larvae
appear in the peripheral blood during the day.

The reason why these larvae appear in the peripheral
circulation only at these times is not yet known. Sleep is
not the only factor which influences them, because the
larvae begin to appear in the peripheral circulation before
the human host is asleep. Further, the microfilarial larvae
of what is apparently the same species of filarial worm do not
show this nightly (nocturnal) periodicity, as it is called,
when they infect the blood of natives of the Philippine
Islands, Fiji, Samoa, Tahiti and some regions near to these
islands. The microfilarial larvae of the related species
Loa loa, which is called the eye-worm, show, moreover, a

daily (diurnal) not a nightly periodicity, appearing in the peripheral blood during the daytime only.

This and some other facts have given rise to the suggestion that there is some relation between the periodicity of the microfilarial larvae and the times when the blood-sucking intermediate hosts suck blood, because the intermediate hosts of Bancroft's filarial worm, which shows nocturnal periodicity, are various species of mosquitoes belonging to the genera *Culex, Aëdes, Anopheles* and *Mansonia*, which bite man and suck his blood by night, while those of the eye-worm (*Loa loa*), which shows diurnal periodicity, are biting tabanid flies belonging to the genus *Chrysops*, which tend to bite man more often during the day. There are, however, facts which indicate that other factors probably also influence the periodicities of microfilarial larvae.

When the intermediate host of Bancroft's filarial worm sucks blood containing the microfilarial larvae, these larvae lose their sheaths within 2–6 hours and migrate through the wall of the intestine of the mosquito to reach the muscles inside its thorax. They reach these in 4–17 hours. In these muscles they grow, cast their skins twice and reach the third larval stage each of which is 1·4–2 mm. long by 18–23 microns broad. They are now infective larvae exactly comparable to the third infective larvae of the sheep stomachworm. Under the best (optimum) conditions the infective larvae may be formed about 10–11 days after they have been sucked up by the mosquito.

They are now ready to infect man again. They migrate, not down the tube formed by the mouthparts of the mosquito through which it sucks blood, but inside its lower lip (*labium*). When the mosquito sucks blood, the infective larvae break out through a membrane at the tip of the mosquito's lower lip, which guides the other mouthparts by which the puncture of the human skin is made (fig. 54). They then penetrate the human skin. Either they penetrate the unbroken skin or they get in through the puncture in it made by the mouthparts of the mosquito. They are therefore, skin-penetrating larvae, like the infective larvae of the hookworms, and they gain entry into the definitive host by their own efforts.

We can therefore say that Bancroft's filarial worm is a roundworm with a life history similar to those of other roundworms except that it has exchanged the exposure of its larval phases to the risks of the world outside the host for the doubtful advantages of a protected life inside an intermediate host into which the first larva must gain entry or perish. The external world has, however, not been abandoned completely, because the infective larva still enters it for a brief period after it has left the mosquito and before it penetrates the human skin; but its entry into it is made at a moment so favourable to its entry into the definitive host that the risks which attend it are very much reduced. The egg, it will be noticed, not only does not enter the world outside the hosts, but produces a larva enclosed in an egg-shell which has become so thin that the larva can probably nourish itself through it with food obtained from the human blood and lymph in which it lives.

Another feature of the life history of Bancroft's filarial worm is the fact that, although it can use only one species of definitive host, namely, man, it can use a variety of species of intermediate hosts, which belong to more than one genus. The human eyeworm, *Loa loa*, on the other hand, is restricted to two species only of intermediate hosts, which belong to one genus only. Their names are *Chrysops dimidiata* and *C. silacea*, and they are two-winged tabanid flies related to the flies which are intermediate hosts of trypanosomes.

Another species of filarial roundworm which can apparently use man only as its definitive host is *Onchocerca volvulus*, which causes another kind of human filariasis (see Chapter 9). It is, so far as we know at present, restricted to intermediate hosts belonging to only one genus of flies, namely, to 'black flies' of the genus *Simulium*. Another species of this genus, *Onchocerca cervicalis*, which uses the midge, *Culicoides nubeculosus*, as its intermediate host, infects horses and cattle, but there is, as yet, no evidence that this species can use man as its definitive host.

The life history of Bancroft's filarial worm concludes our series of types of indirect life histories during which some phase of the parasitic animal enters the external world. Our next and final type of these life histories is that of a species which, like the pork trichina-worm among the species which do not use intermediate hosts, has excluded the external world altogether. This is the human malarial parasite.

Before we consider it, however, let us fix firmly in our minds one important difference between the life histories of the filarial roundworms just considered and those of the malarial parasites of man. This difference is expressed in the diagrams of these life histories, figs. 54 and 55. Reference to fig. 54 will make it clear that man is the definitive host of Bancroft's filarial worm, because the sexual reproduction of this species occurs in man and not in the mosquito, which is the intermediate host. Reference to fig. 55 will, on the other hand, make it clear that the mosquito is the definitive host of the human malarial parasites, because their sexual reproduction occurs in the mosquito. In man they multiply their numbers asexually, so that man is the intermediate host. Man is, in fact, the only species of animal which can act as the intermediate host of the four species of malarial parasites which cause malaria in him, namely, *Plasmodium vivax*, *P. falciparum*, *P. malariae* and *P. ovale*. A fifth species, *P. knowlesi*, which is usually parasitic in the monkey *Macacus rhesus*, will live in human blood if it is experimentally introduced into it.

Type IV B (fig. 55) : *Plasmodium vivax*, the malarial parasite which causes human benign tertian malaria.

The changes which the human malarial parasites undergo during their life histories are very similar to those which the coccidian *Eimeria caviae* undergoes. *Plasmodium vivax*, for example, the cause of benign tertian malaria of man, exhibits, in the course of its life cycle, an alternation of sexual and asexual generations exactly similar to that shown by *Eimeria caviae*. The difference is that both the sexual and the asexual generations of *E. caviae* occur in the same host, the guinea-pig, while those of *Plasmodium vivax* and the other malarial parasites require separate hosts. The asexual generation occurs in the intermediate host, man, while the sexual generation occurs in the definitive host, the anopheline mosquito.

The product of the sexual process in both *Eimeria caviae* and *Plasmodium* is a fertilised egg called a zygote; but, because *Plasmodium* never enters the external world, its zygote does not develop a resistant envelope around itself. It is not enclosed in an oocyst, but is a soft-walled structure which is motile and able to bore its way into the wall of the mosquito's stomach, where it feeds upon the tissue juices of the mosquito. In it we can distinguish masses of protoplasm formed by subdivision of the fertilised female gametocyte. These are exactly comparable to the sporoblasts of *Eimeria caviae* and they produce the infective sporozoites.

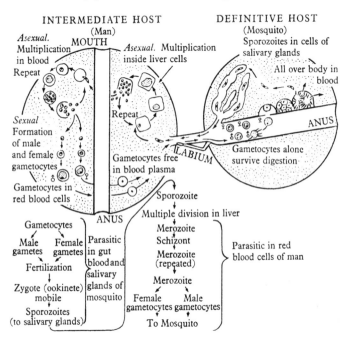

INTERMEDIATE HOST
(Man)

DEFINITIVE HOST
(Mosquito)

Asexual.
Multiplication MOUTH
in blood

Asexual. Multiplication
inside liver cells

Repeat

Sporozoites in cells of
salivary glands

All over body in
blood

Sexual
Formation
of male
and female
gametocytes

Repeat

L A B I U M
Gametocytes free
in blood plasma

ANUS

Gametocytes alone
survive digestion

Gametocytes in
red blood cells

Sporozoite

Gametocytes

ANUS

Multiple division in liver

Male   Female
gametes   gametes

Parasitic
in gut
blood and
salivary
glands of
mosquito

Merozoite

Schizont

Merozoite
(repeated)

Merozoite

Female   Male
gametocytes gametocytes

To Mosquito

Parasitic in red
blood cells of man

Fertilization

Zygote (ookinete)
mobile

Sporozoites
(to salivary glands)

Fig. 55. Diagram of the life history of *Plasmodium vivax*, the cause of human benign tertian malaria, Type IV B

The sporozoites of *Plasmodium vivax*, each of which is about 15 microns long, are liberated into the blood in the body cavity of the mosquito and are carried all over its body. They accumulate especially inside the cells of the mosquito's salivary glands. The mosquito is then infective. It transfers these sporozoites to human blood when next it takes a blood meal. The transfer to the intermediate host is thus accomplished by the definitive host, *Plasmodium* being unable to make this transfer by its own efforts.

Formerly it was believed that the sporozoites, when they reach the blood of the intermediate host (man), immediately enter the red blood cells to feed upon them and grow up inside them. It has recently been shown, however, that, before they enter the red blood cells, they enter the liver of man and there undergo, inside the cells of the liver, a process of multiple division (schizogony) analogous to the schizogony of *Eimeria caviae*. The products of this division (*merozoites*) then enter the red blood cells.

This multiple division in the cells of the liver before the parasites enter the blood cells occurs also during the life histories of *Plasmodium falciparum*, the cause of human malignant tertian malaria, and during that of *P. cynomolgi*, found in the blood of monkeys. Its occurrence has been suspected for many years, but was not proved experimentally until 1948. It was strongly suggested by the proof that, during the life history of *Plasmodium gallinaceum*, parasitic in the red blood cells of birds, a similar multiplication occurs inside certain cells (monocytes) of the liver, spleen, kidneys and blood capillaries, before the products of this multiplication enter the red blood cells to follow a life history essentially the same as that followed by the species of *Plasmodium* which infect man. Phases of malarial parasites which thus develop in cells which are not red blood cells are called *exo-erythrocytic*, or less aptly *pre-erythrocytic*, phases.

Its occurrence during the life histories of the two species of human malarial parasites just mentioned may explain why it is that no sign of malarial parasites can be found in human blood during a period of a week to 10 days, called the incubation period, after the sporozoites have been injected by the mosquito. It may also explain why it is that, if antimalarial drugs are given during this incubation period, they may fail to prevent the subsequent development of malaria, and why it is that blood removed during the incubation period and injected into another person will not communicate malaria to that person.

It is now known that merozoites of *P. cynomolgi* produced by schizogony in the cells of the monkey's liver may invade other liver cells and repeat the pre-erythrocytic schizogony, which may go on while schizogony is also in progress in the red blood cells. It is suggested that malarial parasites may thus persist in the liver after those present in the blood cells have been destroyed, either by the resistance to them acquired by the host (see Chapter 7) or by treatment with antimalarial drugs, and that these parasites persisting in the liver may provide parasites which cause relapses of malaria whose origin has not been otherwise explained.

When the incubation period is over, however, phases of the malarial parasites which resemble the merozoites of *Eimeria caviae* appear in the blood and enter red blood cells. After their entry they first appear in these cells in the form of delicate rings of cytoplasm, each about 2·5 microns in diameter, each with a nucleus at one side and a vacuole full of fluid in the centre. Their appearance has been compared to that of a signet ring, and this phase is often called the signet-ring, or ring, stage.

The ring stage then grows and becomes irregular in shape until it practically fills the red blood cell, which has now become enlarged and pale. This stage is the trophozoite.

The full-grown trophozoite then undergoes asexual multiple division (*schizogony*), forming 12–24 asexually produced, rounded merozoites, each about 1·5 mm. long. These are liberated into the blood, from which they enter other red blood cells.

The liberation of these merozoites and the breakdown of the destroyed red blood cells in which they are formed set free substances which cause the shivering, fever and other symptoms of the malarial 'attack'. It is well known that these attacks occur at regular intervals of time, and we know that the time required by the merozoites of each species of the human malarial parasite to become full-grown trophozoites and to undergo schizogony coincides with the time interval between the typical 'attacks' of malaria. Thus *Plasmodium vivax*, the cause of benign tertian malaria, undergoes schizogony every 48 hours. *P. falciparum*, the cause of malignant tertian malaria, undergoes it every 36–48 hours (although in many human beings a fever caused by this species may be irregular). *P. malariae*, the cause of quartan malaria, undergoes it every 72 hours. *P. ovale* undergoes it every 48 hours. Naturally, if a human being is infected, as he may be, by more than one of these species of *Plasmodium*, or if he is infected by only one species at different times, so that the parasites introduced at each infective bite of the mosquito are at different stages of their development, the different batches of parasites inoculated by the mosquito will undergo schizogony at different times and malarial 'attacks' will not occur regularly every 48 or 72 hours, but may be more frequent and even irregular.

Just as the asexual schizogonous multiplication of individuals of *Eimeria caviae* ultimately ceases and is succeeded by the formation, not of schizonts, but of the male and female gametocytes of the sexual generation, so there comes a time when the merozoites of *Plasmodium* grow up, not into schizonts, but into male and female gametocytes. These gametocytes develop in human red blood cells until they are mature. They are then set free from the cells into the fluid part of the blood. The actual gametes, however, cannot be formed in human blood, nor can fertilisation of the female gamete be effected there. This can occur only in the blood-sucking females of the mosquito intermediate hosts.

When the correct species of mosquito intermediate host sucks up the gametocytes with a meal of blood, its digestive juices digest any asexual phases of *Plasmodium* which may have been sucked up with the gametocytes and the game-tocytes alone survive. They pass to the hinder part of the mosquito's mid-gut (stomach) and there about four or eight male gametes, each about 15–20 mm. long, are formed by each male gametocyte. Meanwhile the nucleus of the female gametocyte undergoes changes which prepare it for fertilisation by the male gamete. The male gamete enters the female gamete, their nuclei fuse and the fertilised egg or *zygote* is thus formed.

This zygote is, as we have already noted, not a passive organism, as the zygote of *Eimeria caviae* is. It elongates and becomes actually mobile and is therefore called an *ookinete*. It penetrates through the lining of the mosquito's mid-gut and comes to rest in the wall of this, where it rounds itself off and a cyst-wall, formed partly by the mosquito and partly by the parasite, develops around it. At this stage it corresponds to the oocyst of *E. caviae*. Each zygote then grows till its diameter is 50–60 microns. Its contents, like those of the oocyst of *E. caviae*, divide into masses of protoplasm called *sporoblasts*, but these do not, as the sporoblasts of *E. caviae* do, form resistant walls around themselves to become spores. Instead, they proceed, in the course of 10 days to 3 weeks, to form the infective *sporozoites* which are liberated into the blood of the mosquito, pass into the cells of its salivary glands and are injected with the saliva into the human intermediate host (cf. Chapter 6).

Now that we have the life histories of Bancroft's filarial roundworm and of the single-celled protozoan, *Plasmodium vivax*, in our minds, it will be helpful to summarise briefly certain important differences between them.

First, Bancroft's filarial worm, like all other species of roundworms, cannot multiply the number of individuals which are derived from each fertilised egg. An increase of the number of individuals present in either its definitive or intermediate hosts can only be effected by means of an increase of the number of fertilised eggs (zygotes) produced. The malarial parasite, on the other hand, can, like *Eimeria caviae* and the flukes and some tapeworms, such as *Taenia multiceps* and *Echinococcus granulosus*, increase the number of individuals which are derived from each zygote by multiplication in the intermediate host. It is, in fact, this asexual multiplication which causes the disease called malaria. The increase of the number of individuals produced by it may be enormous (see Chapter 6).

Secondly, the infective larvae of Bancroft's filarial worm enter the definitive host, man, by their own efforts. This species has become dependent upon the contacts made between its intermediate hosts and its single definitive host, but it retains the power of active entry by its own efforts and it enters the external world for the brief period during which this entry is effected. The malarial parasites, on the other hand, are not only dependent upon the contacts made between the mosquito definitive host and the human intermediate host, but they cannot enter the external world. If they did, they would perish. The active part which they play in their transmission from the definitive to the intermediate host is restricted to the entry of their sporozoites into the cells of the mosquito's salivary glands. Once this has been effected, they are entirely dependent upon the mosquito's habit of injecting, before it sucks up blood, saliva containing *anticoagulins*, which are substances that prevent the clotting of blood. This injection of saliva transmits them into human blood, not by the route taken by the infective larvae of Bancroft's filarial worm through the mosquito's labrum, but through the sucking tube formed by the upper lip and hypopharynx of the mosquito (see Chapter 5).

In addition to this, their sexual phases (*gametocytes*), which pass back from the mosquito to man, are also dependent upon the mosquito's habit of sucking human blood; they also enter man through the sucking tube and enter it passively, being unable to effect entry by their own efforts.

The dependence of the malarial parasites upon the habits of the mosquito and also upon the structure of its mouth-parts and upon its particular method of sucking blood is thus extreme. It has, indeed, gone so far that they are restricted even more closely than the filarial roundworms are to certain species of hosts. The filarial roundworms illustrate restriction to one species only of definitive host, namely, man, combined with varying degrees of increasing restriction of the number of the intermediate hosts in which their larval phases can develop. Some species of them can live in a comparatively large number of species of intermediate host, others in only a few species. The malarial parasites, on the other hand, can live in one species only of intermediate host, namely man, and one genus only of definitive mosquito host, namely, *Anopheles*. Further, the females only of these anopheline mosquitoes can act as definitive hosts of the malarial parasite, because the females only suck the human blood in which the malarial parasites live. Species of anopheline mosquitoes which never suck human blood are therefore excluded. The male mosquitoes are also excluded, because they feed upon the juices of fruits and plants.

This restriction of the number of definitive and intermediate hosts in which the human malarial parasites can develop is a good example of the adaptation of a parasitic animal to a limited number of hosts. It is one of the factors which limit the spread of the human malarial parasites which are discussed at the end of Chapter 10.

CHAPTER 5

SOME EFFECTS OF PARASITIC LIFE
UPON THE PARASITIC ANIMAL, I

Every animal, whatever its mode of life, is gradually altered in various ways by the slow processes of evolution, and the parasitic animal, far from being an exception to this rule, exemplifies it clearly. The modifications which it undergoes after it has become parasitic affect both its structure and its physiological processes, because alterations of structure and function must proceed together. It is difficult, therefore, to discuss structural alterations apart from the physiological modifications which accompany them. We can, however, for descriptive purposes, discuss the various modifications caused by parasitic life under the following headings:

1. Method of feeding.
2. Attachment to the host.
3. Reduction or loss of organs.
4. Modifications of the reproductive processes.
5. Correlations of the life history of the parasitic animal with that of the host.

Not included in this list are certain important and very interesting physiological features of parasitic animals, such as their respiratory adaptations to life inside the bodies of various kinds of animals which are their hosts and their excretory and other metabolic processes, some of which may affect the host considerably. Space is not available for adequate discussions of these features, but some of them are briefly indicated in various parts of this book.

1. Method of Feeding

Alterations of structure which are associated with the method of feeding vary according to the nature of the food

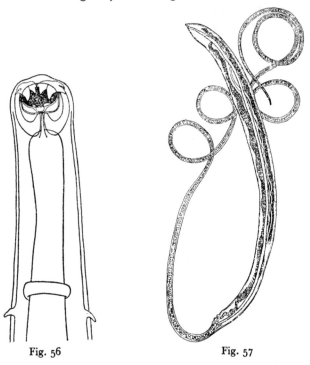

Fig. 56   Fig. 57

Fig. 56. Head and mouth of the human hookworm, *Ancylostoma duodenale*, showing the teeth

Fig. 57. The whipworm of the dog and fox, *Trichuris vulpis*, showing the thread-like anterior portion which penetrates the wall of the caecum for feeding. Total length 45–75 mm.

of the parasitic animal. When it browses on the tissues of its host, it may develop teeth which help it to rasp these tissues off; when it feeds upon liquid food it may develop digestive juices which liquefy the host's tissues; or it may feed upon tissue juices or blood, either through special modifications of the mouth or through sucking tubes which ramify in the host's body. When fluid of any kind is removed from the host, the pharynx of the parasitic animal is often a powerful muscular structure which sucks out the fluid and there may also be extensions of the stomach or of some other part of the food canal in which the food can be stored up. At the same time the function of feeding may be closely associated with the function of attachment to the host, and the two functions may have evolved side by side and may even be performed by the same structures.

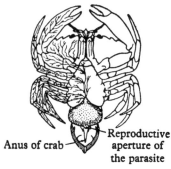

Reproductive aperture of the parasite

Anus of crab

Fig. 58. *Sacculina carcini* (dotted shading) on the abdomen of a crab, showing the root-like processes by means of which it feeds on the crab's tissues

Good examples of modification of the mouth region combined with a sucking pharynx are provided by the blood-sucking insects and the hookworms. The mouth-parts of the female mosquito (fig. 59), for example, are adapted for sucking blood and plant juices. The upper and lower lips are elongated and the mandibles and maxillae are modified to form a delicate stabbing instrument. Together the mouth-parts form a structure called the *proboscis*. The elongated upper lip, which is called the *labrum-epipharynx*, is like a tube which is incomplete on its ventral side. Its shape has been compared to that of the outer cover of a bicycle or motor-car tyre. Along the incomplete ventral surface of it is laid one of the mouth-parts called the *hypopharynx*, so that the two together form a complete tube through which the blood is sucked up. The lower lip, or *labium*, is also elongated and has at its tip two portions

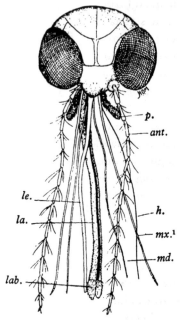

Fig. 59. Head and mouth-parts of a female mosquito, *Culex* sp. *ant.* antennae; *p.* maxillary palp. The palps feel for a place on the skin suitable for a bite; *la.* labium (lower lip), formed from the second maxillae, which provides a groove in which the mandibles (*md.*) and first maxillae (*mx.*[1]) lie. The mandibles and maxillae, guided by the labella (*lab.*), as in figs. 54 and 55, make the wound. Blood is sucked up in a tube formed by the labrum-epipharynx or upper lip (*le.*) and the hypopharynx (*h.*) applied below it

called the *labella*. When the mosquito feeds, it feels over the skin with its maxillary palps to find a suitable soft place for the bite. The skin is then pierced by the pointed tips of the elongated mandibles and maxillae. These are guided between the two labella at the tip of the labium (see figs. 54,

55, 59). As they are thrust into the skin the labium bends near its base to form a loop. The saliva, which contains anticoagulins that delay the clotting of the host's blood, is poured into the blood through an opening at the tip of the hypopharynx. The stylets of the male cannot pierce animals or plants to obtain fluid from them, so that the male must feed upon fluids found elsewhere. These delicate mouth-parts of the mosquito may be compared with those of the tsetse flies (fig. 61) and with the coarser ones of the tabanid flies (fig. 60). The latter inflict a more irritating injury.

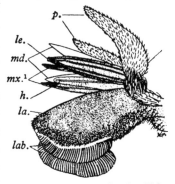

Fig. 60. Mouth-parts of a tabanid fly. Lettering as in fig. 59. The wound is made by the mandibles and first maxillae, but these are coarser and inflict more injury than those of the mosquito do. Blood is sucked up in a tube formed by the upper lip and hypostome, but these flies can also suck in water and other fluids along the channels shown on the oval labella

The mouth-parts of the sucking lice (Siphunculata) show an even more remarkable adaptation to the habit of sucking blood. Those of *Pediculus humanus*, the common louse which infests man and transmits *Rickettsia prowazeki*, the cause of epidemic typhus, illustrate these adaptations well. The mouth-parts of this louse are minute and are, when the insect is not sucking blood, completely withdrawn into the head. They consist of two tubes, one above the other. The upper tube opens upon a short projection called the *rostrum*, upon which there are small teeth and, when the louse is about to suck blood, these teeth maintain a hold upon the host. Muscles then bring the pharynx into contact with the skin of the host.

The lower tube is an extension of the front end of the upper tube, and in it are two stabbers (*stylets*), which perforate the skin of the host when the rostrum has got a hold upon it. The insect then pours saliva into the puncture and the pumping action of the pharynx sucks up blood.

The hookworms, which also suck blood, have developed a bellshaped expansion of the head end (fig. 56), on the

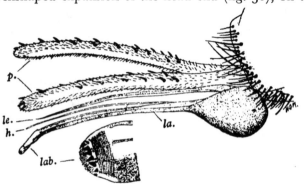

Fig. 61. Mouth-parts of a tsetse fly, *Glossina pallidipes*. Lettering as in fig. 59. Blood is sucked up in a tube formed by the upper lip and hypostome, but the mandibles and first maxillae have been lost. The wound is made by the toothed rasp on the labella (shown in the enlarged figure of these) and the labium is very strong. The mouth-parts of the stable-fly, *Stomoxys calcitrans*, are similar to these

inside walls of which there are teeth. The wide opening of this stiff-walled bell is placed against the soft wall of the host's small intestine, and a plug of the tissues of the intestinal wall is sucked into the bell by the pumping action of the muscles of the pharynx of the worm. The teeth are then used to abrade the tissues of the plug of the host's intestinal wall which has been sucked in.

Gradually the hookworm penetrates into the intestinal wall until it reaches a small artery. It then abrades the wall of this artery until a flow of arterial blood is obtained. From glands in its body the hookworm secretes anticoagulins, which prevent the clotting of the blood of its host, and it then sucks in the prodigious quantities of arterial blood which are mentioned in Chapter 8.

The leeches are also good examples of the development of wounding teeth combined with a sucking apparatus; but the leeches have in addition organs of attachment in the form of suckers and also extensions of the stomach in which the blood obtained is stored up. Most of the leeches maintain a hold upon their hosts by means of two suckers, one at the front end around the mouth, which is a development of the lips, and one at the hinder end behind the vent (*anus*).

The family of leeches called the Gnathobdellidae, to which the medicinal leech, *Hirudo medicinalis* (fig. 6), belongs, possess three lips or 'jaws', which are muscular cushions covered with chitin and beset with eighty or ninety teeth in all. The 'jaws' of the medicinal leech can be moved independently, each in a separate arc, so that the 'bite' of this leech leaves a characteristic triradiate scar on the skin. Along each side of the stomach, which is sometimes called the crop, there are lateral pouches (*caeca*), in which blood can be stored up, so that the leech can do without blood for long periods of time.

The diet of some species of leeches varies, however, at different periods of their development. *H. medicinalis*, for instance, lives, when it is young, on the blood of insects, later upon the blood of frogs and fishes and finally upon the blood of warmblooded animals, when this can be obtained. Various bacteria and other organisms which cause disease, including some trypanosomes, have been found in the blood sucked by this species of leech, and it is believed that *Trypanosoma brucei*, *T. equiperdum* and the bacillus which causes anthrax may be mechanically transmitted (see Chapter 8) by this leech to man. The leech is not a host of these parasites. Two other species of leech which are used medicinally for drawing off blood from man are the Algerian dragon, *Hirudo troctina*, and, in Pakistan, *Gnathobdella ferox*.

The effects of other gnathobdellid leeches, namely, species of the genera *Limnatis* and *Haemadipsa*, are described in Chapter 8.

The method of feeding shown by the gnathobdellid leeches just mentioned may be contrasted with that shown by the leeches belonging to the family Rhynchobdellidae, which possess a muscular proboscis that can be protruded and inserted into the tissues of the host. This proboscis is, in fact, the anterior end of the pharynx, which can be retracted into a temporary sheath and projected again. One or all the jaws of the members of this family may be absent and there are few or no lateral pouches of the crop. This family includes genera, such as *Piscicola*, species of which are intermediate hosts of relatives of the trypanosomes be-

PLATE IV

Fig. 62. Blocking of the small intestine of the pig by numbers of the roundworm, *Ascaris lumbricoides*

Fig. 63. Blocking of the trachea and bronchi of a sheep by numbers of the lungworm, *Dictyocaulus filaria*

Fig. 64. Larvae of the horse bot-fly, *Gastrophilus* sp., feeding on the wall of the stomach of a horse. Length 13–20 mm.

Fig. 65. The large stomachworm of sheep, *Haemonchus contortus*, on the wall of the fourth stomach. Length of worms: female 18–30 mm., male 10–20 mm.

Fig. 66. Liver of a turkey showing the damage done by the flagellate protozoon, *Histomonas meleagridis*, the cause of histomoniasis

Fig. 67. Liver of a sheep infected by the hydatid bladderworms (indicated by arrows) of the dog tapeworm, *Echinococcus granulosus*

PLATE IV

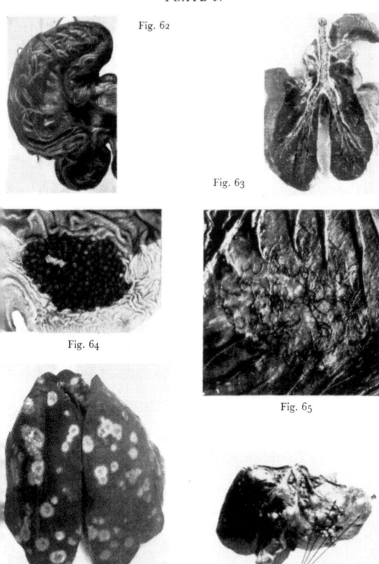

Fig. 62

Fig. 63

Fig. 64

Fig. 65

Fig. 66

Fig. 67

PLATE V

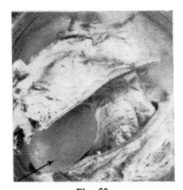

Fig. 68

Fig. 69

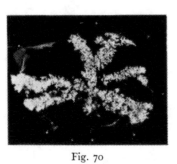

Fig. 70

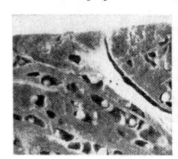

Fig. 71

Fig. 72

Fig. 73

longing to the genus *Cryptobia* (Trypanoplasma), which are parasitic in the blood of freshwater fishes (carp, goldfish), whose blood these leeches suck, and *Pontobdella*, species of which are intermediate hosts of trypanosomes parasitic in the blood of marine fishes, such as the ray.

Still other rhynchobdellid species are parasitic in the mouths of tortoises and crocodiles and upon various Amphibia and some snails. The fresh-water genus *Clepsine* is interesting because some of its species brood over their eggs, which are attached to some foreign body in the water, while the females of two species of it attach their eggs to the ventral surfaces of their bodies, to which their young are also attached after they hatch out.

Among the backboned animals the hagfishes and the few species of bats which suck blood show interesting modifications which enable these species to suck the blood and other tissue juices of their hosts.

The most important of the few species of bats which suck blood are species of the genus *Desmodus*, whose incisor teeth

PLATE V

Fig. 68. Brain of a sheep infected by the coenurus bladderworm (indicated by the arrow) of the tapeworm, *Taenia multiceps*. Diameter of bladderworm up to 5 cm. or more

Fig. 69. The coenurus bladderworm of the tapeworm, *Taenia serialis*, from the abdomen of a rabbit (actual diameter 6 inches)

Fig. 70. Portion of the coenurus bladderworm shown in Fig. 69, opened and more highly magnified to show the numerous tapeworm heads arranged in rows on the wall of the cyst

Fig. 71. Pork infected by the cysticercus bladderworms, *Cysticercus cellulosae*, of the pork tapeworm, *Taenia solium*. Size of bladderworms 5–8 × 6–20 mm.

Fig. 72. Beef infected by the cysticercus bladderworms, *Cysticercus bovis*, of the beef tapeworm, *Taenia saginata*. Size of bladderworms 4–6 × 7–10 mm.

Fig. 73. The larvae of the pork trichina-worm, *Trichinella spiralis*, encysted in pork. Size of each cyst 0·4 × 0·25 mm.

are enlarged and have sharp edges. With them these bats make a shallow, clean-cut scoop from the flesh, which bleeds much longer than a puncture would. This compensates the bat for its lack of the substances which prevent the clotting of the blood which are secreted by the hookworms, leeches and some kinds of blood-sucking insects. The stomach of *Desmodus*, however, like the stomachs of the medicinal leech and its relations, is adapted to accommodate the blood taken in. In *Desmodus* this adaptation takes the form of a long, tube-like extension of the stomach, which holds and digests the blood.

The remarkable cunning with which some kinds of blood-sucking bats stalk their victims and literally steal their blood must also be reckoned among the modifications which their temporarily parasitic habits have produced. Species of *Desmodus* attack cattle, horses and other animals, including man and poultry, when they are asleep at night. They watch their victims carefully and, when they are asleep, they walk or sidle up to them and scoop out a piece of flesh so delicately that the sleeping animal is often not aware of the bite until the bleeding is discovered in the morning. Two other species of blood-sucking bats, belonging to the genera *Diphylla* and *Diaemus*, are rare. The related species, *Vampyrus spectrum*, which used to be regarded as a blood-sucker, lives mainly upon fruit.

The hagfishes belong to the primitive group of fishes called the Cyclostomes, a name which refers to the circular opening into their mouths. All of these fishes have a worm-like shape, and perhaps the best-known example of them is the lamprey. It is known that the lamprey will sometimes attack living fish; but the hagfishes much more frequently attack living fish and are parasitic upon them.

The hagfish has two rows of teeth on its powerful tongue and one median tooth upon the roof of its mouth. Its eyes are very imperfect and are buried beneath the skin, probably because the hagfish burrows deeply into the tissues of the fish which it attacks, so that its eyes have become useless. For the same reason its gill openings are connected by long tubes to a single opening on the surface much farther back than the position of the gill openings of the lamprey, so that the hagfish can breathe water while its head end is buried in the body of the fish upon which it is parasitic.

Some species of hagfish can attach themselves so firmly by means of their suctorial mouths to the living fish that these fish can only rarely shake them off. They then rasp off the flesh of the fish and suck their blood. Some species of them consume the fish's muscles until little is left of the living fish except its bones and viscera and the fish dies. Great damage is done to fisheries in the Monterey region of California by some kinds of hagfishes.

It has been said above that the leeches combine adaptations to their method of feeding with the development of suckers which attach them to their hosts while they are feeding. This introduces us to the variety of ways in which parasitic animals combine these two functions. A comparatively well-known example of their combination is provided by the mouth-parts of the ticks.

The common castor bean or grass tick, *Ixodes ricinus*, for example, which sucks the blood of sheep, cattle, dogs and other mammals, has mouth-parts (figs. 11, 81) which are set upon a basal piece called the *capitulum*, which arises from the anterior end of the tick's body. The mouth-parts themselves consist of a pair of sensory processes called *pedipalps*, and a finger-like process in the middle line which is called the *hypostome*. The hypostome is covered with hooklets which are curved backwards, so that once they have taken hold the tick cannot easily be detached. So firm, indeed, is their hold that forcible detachment of the tick from the host may break off the hypostome and leave it behind in the host. There is also a pair of cutting organs (*chelicerae*), whose outer borders are serrated. With these the tick cuts an opening into which the hypostome is inserted. The host's blood is sucked up along a small tube which lies between the chelicerae and the hypostome. This structure also prevents the regurgitation of the host's blood back into the wound caused by the tick.

Among other parasites whose organs of feeding and attachment are combined are the species of *Sacculina* (fig. 58) and other species of parasitic Crustacea (figs. 104–109), described below. The former send branching processes into the bodies of their hosts, which are used primarily for feeding, but they also help to attach these species to their hosts. This method of feeding is accompanied by the degrees of loss of various organs which are also described below. Some Protozoa send similar processes into the marine animals upon which they are parasitic.

Among examples of the loss of organs which result from parasitic life must be included the partial or complete disappearance of both the food canal itself and of its external openings, the mouth and anus. The tapeworms, for example, have lost the mouth and anus and all traces of a food canal, and their peculiar method of feeding through the surface layers of their bodies is discussed in Chapter 8. The thorny-headed worms (Acanthocephala) (fig. 76), which are classified among the roundworms also have no mouth, anus or food canal, and they feed through the surfaces of their bodies; but they have a powerful, retractile proboscis beset with teeth, which resembles the toothed rostellum of some tapeworms (see below).

Some species of them live in the intestines of ducks, swans and other aquatic birds, others in the intestines of pigs and dogs. The former species in avian hosts belong to the family Echinorhynchidae and are small; some of the latter species belong to the family Gigantorhynchidae and some of these are large. *Macracanthorhynchus hirudinaceus*, for example, may be 10 cm. long. The infection of man with this species has been recorded, but it is rare.

Finally we may mention another interesting method of feeding exemplified by the nematode whipworms, which are relatives of the pork trichina-worm. An example of them is *Trichuris vulpis* (fig. 57), parasitic in the caecum and other parts of the intestine of the dog and fox. Another species, *T. trichiura*, is parasitic in the caecum and colon of man, some monkeys and the pig, where it does comparatively little harm. The shape of these worms recalls that of a whip and its handle, because the anterior two-thirds or so of their bodies is narrowed to form a hair-like tube which is buried in the lining of the host's intestine. The worms secrete digestive juices which reduce the host's tissues to a fluid which the worms suck in. Some other parasitic roundworms or their larvae feed in this way, among them being the larvae of the threadworm of the horse, *Oxyuris equi*.

2. METHODS OF ATTACHMENT TO THE HOST

Apart from the organs just described which combine the functions of attachment and nutrition, there are many which are organs of attachment only. Among the simplest of these are the protoplasmic processes which some kinds of parasitic Protozoa thrust into the bodies of their hosts. Processes of this kind are developed by the Protozoa which are called *gregarines*, and they are called *epimerites*. These may be unbranched or they may have branches shaped like hooks or anchors with two or more processes; or they may be root-like extensions into the host's body.

Another group of Protozoa, the Cnidosporidia, make spores which shoot out, under certain conditions, long threads which attach the spores to the walls of the intestine of the fish which are hosts of these species. These threads are shot out with incredible speed and so suddenly that it is hardly possible for the human eye to react quickly enough to photograph them through a microscope. They seem to be actuated by changes in the osmotic pressure in the spore.

Hooklets are developed by many kinds of parasitic animals either for their attachment to the exterior of the host or to its internal organs. Even the eggs of some species may have hooklets and similar adaptations which give them a hold upon the structures upon which they are laid.

Each egg of the warble-flies belonging to the genus *Hypoderma* (figs. 83, 84) is set upon a short stalk, which has a grooved, elliptical base which fits the hairs of the cattle that are the hosts of these species, and each egg is fixed to the hair by a secretion made by the female fly, which hardens like a cement. These eggs are fixed near the roots of the hairs. The operculated eggs of the horse bot-flies *Gastrophilus intestinalis* and *G. nasalis* (fig. 74), on the other hand, are without stalks and are fixed only by a kind of cement; but those of *G. haemorrhoidalis* (fig. 74) have a stalk laid along the hair. The eggs of bot-flies are fixed nearer the free ends of the hairs.

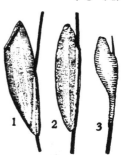

The unstalked, operculated eggs (nits) of biting and sucking lice (figs. 85, 86) are cemented to the feathers or hairs of the hosts of these insects. The human head-lice attach their eggs (fig. 85) to the hair of the human head, and the human body-lice attach theirs to the fibres of the clothing, especially along the seams and where the fibres cross. The attachment of the eggs of the tropical warble-fly, *Dermatobia hominis*, to the insects which transport these eggs to the animals in which the larvae that hatch from them are parasitic, is

Fig. 74. Transversely striated eggs of horse bot-flies attached to hairs. 1. Sessile egg of *Gastrophilus intestinalis*, Common Bot-fly. Length 1·3 mm. 2. Sessile egg of *G. nasalis*, Throat Bot-fly. Length 1·2 mm. 3. Stalked egg of *G. haemorrhoidalis*, Lip Bot-fly. Length, including the stalk, 1·4 mm.

described in Chapter 9. The function of the stickiness of the eggs of the human seat-worm, *Enterobius vermicularis*, is also discussed in Chapter 9.

Adult parasitic animals show us many examples of hooks and similar structures. They are well developed by many species which live on the exterior of their hosts and must maintain a hold there. But, just as modifications of the feeding organs may be evolved along with development of means of attachment to the host, so the development of hooks and similar contrivances may be associated with other modifications, such as the loss of wings and of the locomotor

126

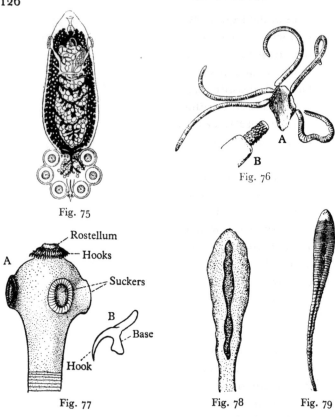

Fig. 75

Fig. 76

Fig. 77

Fig. 78

Fig. 79

Fig. 75. The fluke, *Polystoma integerrimum*, showing its six cup-like suckers at the hind end, reinforced by a pair of hooks. Length 2 mm.

Fig. 76. Five specimens (A) of a thorny-headed roundworm, *Echinorhynchus gadi* (*acus*), attached to the intestine of the whiting by the spiny proboscis shown in (B). Female 45–80 mm., male 20 mm. long

Fig. 77. Head (A) of the pork tapeworm, *Taenia solium*, showing its suckers, hooks and young segments forming at the neck. B, one of the hooks enlarged. Diameter of the globular head, 0·6–1 mm.

Fig. 78. Head of the broad tapeworm of man, *Diphyllobothrium latum*, showing one of the two grooves (*bothridia*) on its head, which act as suckers. Size of the head 2·5 × 1 mm.

Fig. 79. The 'tongueworm', *Linguatula serrata*. Female 8–13 cm., male 1·8–2 cm. long

functions of the legs. They may also be associated with the development of means of evading the efforts of the host to dislodge the parasitic animal (cf. Chapter 10).

The correlation of all these modifications is well shown by some parasitic insects, such as the fleas, the lice and the sheep ked. These three differ-
ent kinds of insects have all lost their wings, which they no longer require. All of them have developed claws on their feet, by means of which they cling to their hosts. The fleas (fig. 89) have two claws at the end of each of their three pairs of legs, the hindmost pair of legs is specially large and gives the flea its well-known jump-ing powers, and their bodies are flattened from side to side and are difficult to hold even when they are caught, be-cause the chitin of which the

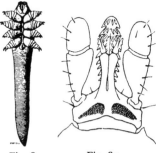

Fig. 80   Fig. 81

Fig. 80. The hair-follicle mite, *De-modex folliculorum*. Female 0·36 mm., male 0·3 mm. long

Fig. 81. The rostrum of the tick, *Ixodes ricinus*, showing, in the centre, the toothed hypostome and cheli-cerae (mandibles), and, at each side, a pedipalp

cuticle is formed is smooth and slippery, and because the bristles on the segments are directed backwards. The host therefore, as many a human being has found, may find it difficult to get rid of a flea. The flea's antennae are shortened and are tucked away in depressions in the side of the head, but they can be rotated out of these.

Other features of the fleas, such as the mechanism which prevents regurgitation of the contents of the stomach while the pharynx is sucking blood, play an important part in the spread of diseases, such as plague, which are transmitted by some species of fleas. These modifications are, however, outside the scope of this book; they are well described in text-books of parasitology.

The bed-bug, *Cimex lectularius* (fig. 10), the bites of which may cause serious inconvenience to human beings at night, has two very sharp claws at the end of each of its six legs. The lice (figs. 90–99), unlike the fleas, have bodies which are flattened from above downwards (dorso-ventrally). Both the biting and the sucking lice have two claws on the ends of each of their three pairs of legs, with the exception of two genera of biting lice, namely, *Trichodectes* and *Gyropus*. To the former genus belongs the biting louse of the dog, *Trichodectes canis*.

The sheep ked, *Melophagus ovinus* (fig. 88), is not a louse, but a wingless member of the section of the dipterous insects called the Pupipara, all of which live parasitically either on the surface of their hosts or in their surface layers and show remarkable adaptations to parasitic life (cf. § 3 below).

PLATE VI

Fig. 82. Stalked egg of the warble-fly, *Hypoderma bovis*, attached singly to a hair of a cow. Length about 1 mm.

Fig. 83. Cercariae of the liver fluke, *Fasciola hepatica*, encysted on grass blades. They appear as small white specks. Size 0·3 × 0·2 mm.

Fig. 84. Stalked eggs of the warble-fly, *Hypoderma lineatum*, attached in rows to a hair of a cow. Length of each egg about 1 mm.

Fig. 85. Sessile egg of the human head-louse, *Pediculus humanus* var. *capitis*, attached to a hair. Length rather less than 1 mm.

Fig. 86. Sessile egg of the short-headed sucking louse of cattle, *Haematopinus eurysternus*, attached to the hairs. Size 1·09 × 0·51 mm.

Fig. 87. The mite which causes human scabies, *Sarcoptes scabiei*, showing stalked suckers on the feet (from a wax model). Size, female 0·3 × 0·26 mm., male 0·2 × 0·16 mm.

PLATE VI

Fig. 82

Fig. 83

Fig. 84

Fig. 85

Fig. 86

Fig. 87

PLATE VII

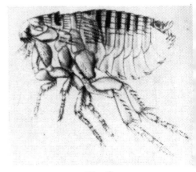

Fig. 89

Fig. 88

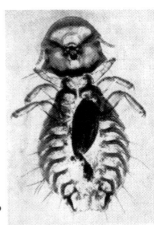

Fig. 90

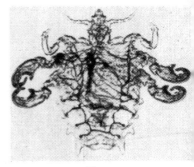

Fig. 91

Fig. 92

The sheep ked lives in the wool of sheep and is sometimes wrongly called a sheep tick. Like the tick it possesses claws on the ends of its legs, but the shape of these claws is different from that of the claws on the tick's legs. The sheep ked can be easily distinguished from a tick if we remember that it is an insect and that the bodies of insects are divided into three main parts, the head, the thorax and the abdomen, while the bodies of ticks consist of one unsegmented piece only, which represents the fused thorax and abdomen. The first few segments of the body form the organs described above which attach the tick to its host. In addition to this feature, the sheep ked has, like all other insects, three pairs of legs, while the tick has four pairs. This feature will therefore distinguish between the adult insect and the adult tick. It will not, however, distinguish between the insect and the larva of the tick, because the latter, like the insect, possesses three pairs of legs. Reference to the number of parts of the whole body will, however, serve to distinguish between these two kinds of parasitic animal.

Many species of mites, which are near relatives of the ticks, bear, on the ends of each of their four pairs of legs, a pair of claws, which fit exactly round the hairs of their

---

PLATE VII

Fig. 88. The sheep ked, *Melophagus ovinus*. Length 4–6 mm.

Fig. 89. The dog-flea, *Ctenocephalides canis*. Length 2–4 mm.

Fig. 90. The brown biting louse of birds, *Goniodes dissimilis*. Female 2·6 mm., male 1·95 mm. long

Fig. 91. The human head-louse, *Pediculus humanus* var. *capitis*, showing hooks on the feet. Female 2·7 mm., male 1·6 mm. long. The human body-louse is rather bigger

Fig. 92. The human crab-louse, *Phthirius pubis*, showing the hooks on the feet. Female 1·5 mm., male 1 mm. long

hosts. *Trombicula autumnalis*, for example, the harvest mite, has these. Only the six-legged larvae of this species suck blood. Species of *Trombicula* transmit the species of the genus *Rickettsia* which cause mite typhus (scrub typhus) of man.

The mites which cause human scabies (fig. 87) and mange of some animals (figs. 13–16), on the other hand, possess upon some of their legs, but not upon all of them, not hooklets, but suckers set upon stalks.

Sarcoptes scabiei (fig. 87) is the cause of human scabies. The female mite is found in a tunnel in the skin, usually at the blind end of it. She lives 4 or 5 weeks, laying during this time 40–50 eggs in the tunnel from which, after 3–5 days, six-legged larvae emerge. These larvae either make lateral tunnels and live in these or they escape from the tunnel of the mother and enter new hair pits (follicles) or make new tunnels. After 2 or 3 days they moult and pass through the two nymphal phases. Eight to fifteen days after the eggs from which they developed were

---

PLATE VIII

Fig. 93. The sucking louse of the pig, *Haematopinus suis*. Female 4·5–6 mm., male 3·5–4·75 mm. long

Fig. 94. Male short-headed sucking louse of cattle, *Haematopinus eurysternus*. Female 3·5–4·75 mm., male 2–3·5 mm. long

Fig. 95. Female long-headed sucking louse of cattle, *Linognathus vituli*. Female 2·25–3 mm., male 2–2·5 mm. long

Fig. 96. The sucking body-louse (blue louse) of sheep, *Linognathus ovillus*. Female 2·5 mm., male 2·25 mm. long

Fig. 97. The sucking foot-louse of sheep, *Linognathus pedalis*. Female 2 mm., male 1·5–1·75 mm. long

Fig. 98. Male small blue sucking louse of cattle, *Solenopotes capillatus*. Female 1·75 mm., male 1·25 mm. long

Fig. 99. The biting louse of sheep, *Bovicola ovis*. Female 1·4–1·8 mm., male 1·2–1·4 mm.

PLATE VIII

Fig. 93

Fig. 94

Fig. 95

Fig. 96

Fig. 98

Fig. 97

Fig. 99

laid they have become adult fertilised females. The males probably live for 4 or 5 weeks.

The mites which cause sarcoptic mange of horses, dogs, cattle, pigs and other animals are usually described as varieties of the species *S. scabiei*. Some of them are transmissible to more than one species of host and some, notably those which normally attack horses and pigs, may live temporarily in the skin of man and may cause a troublesome irritation of it. Other kinds of mange of domesticated animals, and of rabbits, are caused by species of the genera *Psoroptes* (figs. 13–16), *Notoedres*, *Chorioptes* and *Otodectes*, while two species of the genus *Cnemidocoptes* attack poultry, causing the diseases called 'scaly leg' (*C. mutans*) and 'depluming itch' (*C. gallinae*).

The adaptations of the burrowing species of mites to their modes of life are remarkably efficient. Among the structural adaptations of some species of them is the reduction of the legs to stumpy projections. The legs of the species of the genus *Cnemidocoptes gallinae*, for example, the cause of 'depluming itch', a species which burrows into the skin of poultry alongside the shafts of the feathers, with the result that irritation and inflammation occur and the feathers break off and are pulled out by the birds, are short and stumpy, like those of *Demodex folliculorum* (fig. 80), the cause of demodectic mange of man and various mammals. This elongate species of mite is 0·25 mm. long and lives in the hair follicles and also in the sebaceous glands which secrete the waxy sebum, which lubricates the hairs. The opening of the gland may become plugged up by the waxy secretion, so that the condition sometimes called a 'blackhead' develops. The skin may also become thickened and scaly, and the hairs may fall out. If bacterial infection occurs, small abscesses may develop. The elongate body of this species and its short stumpy legs recall the similar features of the tongue-worm of the dog, *Linguatula serrata* (fig. 79), mentioned in Chapter 9, which is not a worm, but is, like the mites, related to the scorpions and spiders (*Arachnida*). It has, in the course of its parasitic life, lost most of the external features of its relatives and attaches itself by means of hooklets on its short, stumpy legs to the lining of the nasal passages of the carnivorous mammals which are the hosts of its adult phase.

Hooklets are also characteristic of many species of tapeworms, which combine their action with that of the suckers present on their heads. Both the hooklets and the suckers of tapeworms show several variations. The hooklets may be set upon a retractile projection of the head called a *rostellum*. This arrangement is shown by the dog tapeworm, *Dipylidium caninum*. Or they may be arranged in one or several rows round the head. The head (fig. 77) of the human pork tapeworm, *Taenia solium*, for example, has 22–32 hooklets in all, set in two rows. Other species of the genus *Taenia* have more or fewer hooklets, or none at all.

The suckers of tapeworms have various shapes. They may be circular or oval in outline or may take more elaborate forms. When both suckers and hooklets are present, the tapeworm is said to be *armed*. The human pork tapeworm, *T. solium*, is an armed species. When there are no hooklets and suckers only are present, the tapeworm is said to be *unarmed*. The human beef tapeworm, *T. saginata*, is unarmed, and another unarmed human species is the fish tapeworm, *Diphyllobothrium latum*, whose suckers have the form of two shallow grooves in the head (fig. 78). It has already been explained that the embryos of tapeworms inside their eggs are called *hexacanth embryos* (fig. 46), because they also possess six hooklets, by means of which they attach themselves to, or tear their way through, the tissues of their intermediate hosts.

The combination of hooks and suckers shown by the armed tapeworms is, however, much more highly developed by the group of flukes called the Monogenea, which live on the surfaces of their hosts and do not require an intermediate host. The hinder ends of these flukes are modified to form adhesive organs, which are sometimes called *opisthaptors*. The opisthaptor usually has hooks, and sometimes suckers or pincer-like clamps as well. In addition to the opisthaptor or posterior sucker, these flukes may have, near the anterior end of the body, a muscular or glandular organ, which is sometimes called the *prohaptor*. The structure of these adhesive organs and the number of them possessed by various species of these flukes varies so much that they cannot be described here. The life history of a member of this group of flukes, *Polystoma integerrimum* (fig. 75), is described in Chapter 6.

The other group of flukes, called the Digenea, which live inside the bodies of their hosts and require an intermediate host, usually have two suckers, one around the mouth and one near the middle of the ventral surface of the body or farther back. The liver fluke of sheep and cattle, *Fasciola hepatica* (fig. 1), whose life history is described in Chapter 4, is a member of this group.

A primitive form of sucker is developed by the flagellate protozoon *Giardia intestinalis*, which lives in the duodenum of man and has on its under-side a circular depressed area with a raised margin, which is applied to the surface of the cells lining this part of the food canal and enables this small

single-celled animal to maintain a hold against the movements of the intestine and the currents in it. Among organs
of attachment whose function is similar to that of hooks and
suckers just described are the adhesive pads developed by
the feet of many insects, such as the forest flies mentioned
below, and the stalked or unstalked suckers on some of
the legs of parasitic mites (figs. 13–16, 87). The parasitic
larvae of the sheep nasal-fly have two powerful hooks by
means of which these larvae attach themselves to the lining
of the nasal passage, in which they are parasitic.

Another interesting form of hooklet is developed by the
glochidium larva (fig. 102) of the bivalve fresh-water swan
mussel, *Anodonta cygnea*. The adult swan mussel is, of course,
not parasitic, but its young larva is. In this respect its life
history resembles the warble-fly.

The shell of the parasitic larva of the swan mussel consists of two triangular halves, hinged along their bases, but
drawn out at the tips of the triangles into two toothed hooks,
one on each half of the shell. This larva does not possess the
muscular foot by means of which the adult swan mussel
ploughs its way through sand or mud. The foot is replaced in
the larva by a gland which secretes a sticky thread comparable to the threads by which the adult sea mussel, *Mytilus
edulis*, fixes itself to rocks, piers, ships and other objects.
When a fresh-water fish, such as a stickleback, passes near
to the glochidium larva of the swan mussel, the larva flaps
the two valves of its shell and this action drives out the
sticky thread. If this thread touches the fish, it sticks to
it, the glochidium larva is brought against the body of the
fish and the hooks of the shell of the larva fix themselves
in the fish's skin. The reaction of the tissues of the fish
take the form of an inflammation, the result of which is the

enclosure of the glochidium in the tissues of the fish. There the glochidial larva lives for some months, feeding upon the fish's tissues and being carried about by its host, so that the species may thus be spread over wide distances. This means of dispersal compensates the swan mussel for the slow movements of the adult mussel, which cannot, in the course of its life, progress very far. Its larvae are also transported in this way farther than they could be transported by the currents of the ponds and streams in which they live. The glochidia are truly parasitic in the skin of the fish and in this respect differ from such developmental phases as the eggs of the human warble-fly (see Chapter 9), which are not parasitic on the insects which disperse them in a similar manner.

3. REDUCTION OR LOSS OF ORGANS

The typical parasitic animal lives in close contact with, or surrounded by, an abundance of food supplied by its host. One of its main problems, therefore, is the establishment of contact with this food supply. This is probably one of the reasons why those phases of the life history which enter the host are often, but not always, motile and able to move about inside the host's body until they reach those tissues of the host in which the sexually mature phases live.

In many instances, however, the adult phases, and some larval phases also, do not move much about the host's body, because they are surrounded by food and can obtain it without the help of locomotive organs. They tend to lead a sedentary life and to undergo the modifications to which this mode of life leads. Among non-parasitic animals a sedentary mode of life tends to produce radial symmetry and to cause reduction or loss of locomotor organs. Parasitic animals do not, as a matter of fact, develop radial symmetry; but they do show us striking examples of the loss or reduction of locomotor organs.

It must not be assumed, however, that the reduction or loss of locomotor organs always deprives the parasitic animal of all power of movement. The muscles of the body may still be efficient. Indeed, they may be so altered that the animal can move very vigorously in spite of the loss of its locomotor organs. Thus a legless dipterous maggot, living on decaying organic matter, can move very actively without any legs; and the peculiar muscles of roundworms, working in a body rendered turgid by fluid inside it, enable these limbless worms to move in fluid or semi-fluid surroundings.

The tapeworm attached to the wall of its host's intestine can move to such an extent that some of the abdominal discomfort caused by its presence there is due to these movements. These movements are, indeed, essential for the approximation of the genital pores of one segment to those of another, so that this hermaphrodite species may fertilise itself. Species of tapeworms which provoke in the host a reaction which prevents the establishment of more than one individual in any one host (see Chapter 7), such as *Taenia saginata*, the beef tapeworm of man, depend entirely upon self-fertilisation effected in this manner. The muscular movements of the detached segments of some species of tapeworms have been described in Chapter 4.

Rather like a worm in appearance, but in reality a relative of the ticks and mites, is the tongueworm, *Linguatula serrata* (fig. 79). It has, as a result of parasitic life, lost the external features of its biological relatives. But more remarkable examples of the reduction or loss of organs caused by parasitic life are provided by some insects and parasitic Crustacea.

Among the insects, the group of Diptera or two-winged flies contains the section called the Pupipara, which includes a number of species which show variable degrees of the loss of wings. All the species of the Pupipara, excepting those of the genus *Braula*, live on the surfaces or in the surface layers of mammals and birds, whose blood they suck. They do not attack man. As the name of the group implies, the larvae of all of them, excepting those of the genus *Braula*, are nourished inside the uteri of the female flies. They are then laid on the ground or in the places where their hosts live, and change almost at once into pupae. The early part of their lives is therefore protected, as that of the vertebrate mammal is, inside the body of the mother. Production of young in an advanced stage of development is characteristic of other parasitic animals and has resulted, no doubt, from the necessity of protecting the young and ensuring their contact with the host. The adult phases of the Pupipara show various degrees of reduction of the wings, correlated with modifications of the legs, of which the following examples may be given.

The family Hippoboscidae are dorso-ventrally flattened flies which are tough and leathery, and the head is sunk into the front of the thorax. These features are adaptations to their life on the surface of their hosts. *Hippobosca equina*, the forest-fly, which attacks horses and cattle, is common in southern England and in Europe. The adult fly is a blood-sucker, and it has strong, toothed claws on its feet which enable it to hold on to the hairs of its hosts. Adhesive pads on the feet also help it to keep a hold. It crawls over the host's body as well and may thus cause much irritation. It has, however, retained its wings, although these are relatively narrow.

An allied species, *Ornithomyia auricularia*, which lives on birds, has even narrower wings, while the wings of *Lipopterna hirudinis*, which lives on swallows and may be found in their nests, have been reduced to strap-like vestiges; and the females of *L. cervi*, which live upon deer, cast off their wings when they find a host. Finally, *Melophagus ovinus* (fig. 88), the sheep ked, which lives in the wool of sheep, has lost its wings, but its feet are well adapted in the manner described above for clinging to the wool of the host.

Members of the family Streblidae are almost all parasitic upon bats, but many species of them possess well-developed wings. The eyes are either vestigial or absent. They possess, on the underside of the flattened head, comb-like organs called *ctenidia* which enable them to hold on to the host. Ctenidia are also possessed by some species of fleas and by bugs belonging to the family Polyctenidae, which also infest bats, living in their fur.

A genus of the Streblidae, both sexes of which are winged, is *Ascodipteron*, which is parasitic upon long-fingered bats belonging to the genus *Miniopterus* in Queensland, the Philippine Islands and the tropics of the Far East. The adult male fly is little altered, but the female has a large proboscis provided at its tip with four series of small chitinous blades. With these the female cuts a small hole in the skin behind the ear of the bat and enters this head first. The wings are then cast off and the legs also are shed, with the exception of their basal joints. The abdomen of the female then enlarges and the head and thorax are telescoped into a pit in the anterior end of the abdomen. The female then lives in a small swelling in the bat's skin, feeding upon the host's tissues and breathing by means of the four spiracles upon the top of the abdomen, which protrudes out of the swelling. The whole female is about 5 mm. long and the uterus contains one larva only, which is nourished in the uterus until it is laid as a pupa. The males live a non-parasitic life.

There is also a flea, *Rhyncopsylla pulex*, which burrows into the skin of the Columbian bat, *Molossus obscurus*, and casts off its limbs. Somewhat similar adaptations are shown by the viviparous earwig, *Arixenia esau*, which lives in the shallow pouch of skin under the wing of the Sarawak bat, *Cheiromoles torquatus*, and has lost its wings and acquired a flattened body adapted to this mode of life. Another species of this genus, *A. jacobsoni*, is found in Java.

The pupiparous family Nycteribidae are also parasitic upon bats and use no other kind of host. They are all wing-less. Many of them are restricted to certain genera or even to certain species of bats. One species occurs upon British bats. The bodies of these insects are flattened. The head is minute and folded back into a groove on the upper side of the thorax.

The family Braulidae contains only the genus *Braula*. The species of this genus are minute and wingless. Unlike other Pupipara, they lay eggs. They are not strictly speaking parasitic, because the eggs are laid among the brood combs of the beehive, and the larvae which emerge from them feed upon the food brought by the bees for their own larvae.

Reduction of the locomotor organs and their adaptation to the function of holding on to the host is also well illustrated by some kinds of parasitic Crustacea which

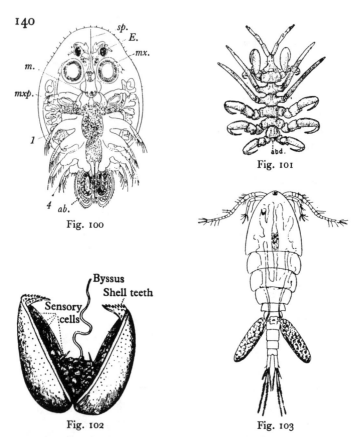

Fig. 100

Fig. 101

Fig. 102

Fig. 103

Fig. 100. A young male 'carp-louse', *Argulus foliaceus*. Female 6–7 mm. long, male smaller. *m.* mandibles and first maxillae, which penetrate into the hosts (fish); *mx.* second maxillae modified to form the rounded suckers shown; *mxp.* hooked maxillipedes, which help the suckers to hold on to the host; *ab.* the small abdomen, with the oval testis in it; *E.* compound eye; *sp.* poison-spine (sting); *1, 4,* two of the eight swimming legs

Fig. 101. A 'whale-louse', *Paracyamus boopis*, parasitic on the skin of the hump-backed whale, *Megaptera boops*. Note the claws on the last three pairs of legs and the very small abdomen (*abd.*). Female 8 mm., male 12 mm. long

Fig. 102. Glochidium larva of the swan mussel, *Anodonta cygnea*, showing the hooks which attach it to the skin of fishes. Size about 0·3 × 0·29 mm.

Fig. 103. Female of the non-parasitic copepod crustacean *Cyclops* sp., intermediate host of the guinea-worm, *Dracunculus medinensis*, and of the broad tapeworm of man, *Diphyllobothrium latum*. The elongate-oval egg sacs are shown in front of the forked tail. Length 1·6–2·4 mm.

show a variety of degrees of alteration of their body-shape and of their limbs due to parasitic life. Some of them have lost all resemblance to their non-parasitic relatives or are so altered that they do not look like Crustacea at all. The occurrence among them of minute males, called dwarf males, which are attached to the bodies of the females, is considered in Chapter 6.

Among the least modified of these parasitic Crustacea are the 'whale-lice' and 'carp-lice'. The 'whale-lice' are amphipods related to the sandhoppers, and an example of them is *Paracyamus boopis* (fig. 101), parasitic on the skin of the hump-backed whale. Like the 'carp-lice', the 'whale-lice' have broad, flattened bodies, but their mouthparts are adapted for biting rather than for sucking the tissue fluids of their hosts and they bite out pits in the skin of their whale hosts and live in these. Clusters of hundreds of them may be found living on the skin of whales. They have, on their last three pairs of legs, claws by means of which they hold on to their hosts.

The 'carp-lice', which are copepods related to the non-parasitic genus *Cyclops* (fig. 103), are found on the surfaces of fresh-water fishes or in their gill chambers or they may be found swimming freely in the water. An example of them is *Argulus foliaceus* (fig. 100), parasitic upon carp and sticklebacks. Their bodies, like those of the 'whale-lice', are broad and flattened, and in this respect they resemble the true (insect) lice which infest the skins of mammals and birds. They suck the blood and tissue fluids of their fish hosts through a tube containing the mandibles and first maxillae, which are modified for penetration of the host. In this respect they differ from the 'whale-lice' and many of the parasitic copepods (see

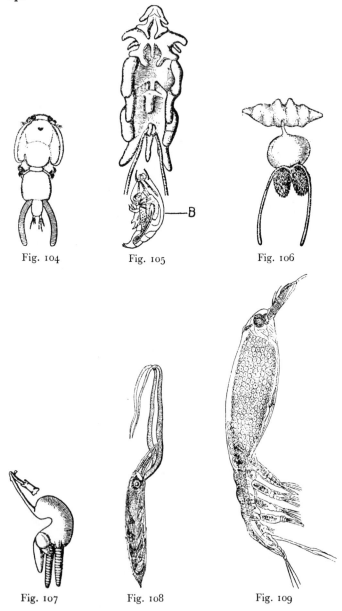

Fig. 104

Fig. 105

Fig. 106

Fig. 107

Fig. 108

Fig. 109

below). The second maxillae are modified to form suckers
and with these and the hooked maxillipedes, the 'carp-
lice' cling to their hosts. They also possess a kind of sting.
Among the other parasitic copepods comparatively
little altered by parasitic life are the 'fish-lice' (Caligidae),
of which *Caligus rapax* (fig. 104) is an example. The bodies
of Caligidae are flat and broad. Both sexes are parasitic
on the skins of fishes, whose blood they suck through a
tube containing the mandibles. Like the 'carp-lice' they
have retained the power of swimming apart from the host.
The second antennae, second maxillae and the maxillipedes are modified as hooks for attachment to the host and
Caligus has, in addition, a pair of suckers at the bases of
the first antennae.

In comparison with the species just described, other
groups of parasitic Crustacea (figs. 105–109) are altered
to a much greater degree. Although the adult females,
when they are fertilised, are recognisable as Crustacea,
they become fixed, after fertilisation, to their fish and other
aquatic hosts, and then the shapes of their bodies are so
greatly altered that it is difficult or impossible to tell from

FIGS. 104–109

Figs. 104–109. Types of parasitic copepod crustacea, showing various
degrees of reduction or loss of organs caused by parasitic life. Fig. 104.
Caligus rapax. The egg sacs are shaded transversely. Female 5–6 mm., male
4–5 mm. Fig. 105. *Chondracanthus lophii.* The dwarf male (B) is shown
below the female, to which it is normally attached. Female 12 mm. long,
male very small. Fig. 106. *Sphyrion (Lesteira) laevigatum.* Female. Female
without egg-strings, 70 mm. long, male not known. Fig. 107. *Haemobaphes
(Lernaea) cyclopterina.* Female. Female about 12 mm. long. Fig. 108. *Cymbasoma rigidum (Haemocera danae).* Full-grown parasitic larva, showing its
feeding processes and the non-parasitic adult (see fig. 109) inside it. Fig. 109.
Cymbasoma rigidum (Haemocera danae). Non-parasitic adult female. Female
2·2–2·5 mm., male 1·5–1·7·5 mm. long

their external appearance that they are Crustacea. Their true relationships are determined only by careful examination of them and by the fact that a true copepod nauplius larva appears in the course of their life histories.

The females of the family Chondracanthidae, for instance, are parasitic on the gills and other soft parts of bony fishes, such as the plaice, sole, skate and angler-fish. The males are small and are attached to the bodies of the females. They are called *dwarf males*. *Chondracanthus lophii* (fig. 105) is an example of this family. The female of *C. lophii* has a soft, unsegmented body produced into irregular lobes. Two forked lobes on the under-side represent altered swimming feet. The minute dwarf male (fig. 105 B) is attached to the hinder end of the female near the two thread-like egg masses. *Sphyrion* (*Lesteira*) *laevigatum* (fig. 106), another member of the Chondracanthidae, is parasitic on the fish *Genypterus blacodes*, a relative of the blennie fishes. It is attached between the skin and the muscles of its host by means of the swollen and transversely elongated head shown in the figure.

Analogous changes are shown by the family Lernaeidae, of which *Haemobaphes* (*Lernaea*) *cyclopterina* is an example. Similar in appearance to this species is *Lernaea branchialis*, which is parasitic on the gills of fishes of the cod and haddock family (Gadidae). The body of the female *Lernaea*, when she becomes parasitic, loses all resemblance to the Crustacea. It becomes reddish and worm-like and grows to a length, excluding the egg-strings, of 40 mm. or so. The head is buried in the gills of the host and is joined to the soft body by a neck. From the head three branched processes extend into the host's gills, which become ulcerated. The maxillae are modified for piercing the gills of the host. The limbs are reduced to microscopic vestiges and the egg sacs are very large. The whole animal is, in fact, reduced to little more than the organs required for feeding, holding on to the host and reproducing the species. The nauplius larva of *L. branchialis*, which reveals its crustacean relationships, is parasitic, not upon fishes of the cod and haddock family, upon which the adults are parasitic, but upon the gills of plaice and its relatives. The males are not attached to the females, but live a non-parasitic life.

Related to the two families just mentioned are the Lernaeopodidae, members of which are attached to their hosts by two arm-like maxillipedes, which either coalesce or are joined together at their tips. They end in front of the head in a chitinous process or in a pad, which is buried in the host's tissues.

Cymbasoma rigidum (*Haemocera danae*) (figs. 108, 109), which is an example of the family Monstrillidae, has a remarkable life history which resembles in one respect the life histories of the insect warble-flies and the nematode Mermithidae. The larval phases of it only are parasitic, the adult males and females living a non-parasitic life. The nauplius larva attaches itself by means of hooks on its third pair of limbs to the skin of marine worms belonging to the group of Polychaeta, to which the lugworm belongs. The nauplius then bores its way into the blood vessels of the worm and casts off its skin and limbs, so that it consists of little more than the degenerating eye of the nauplius and a mass of cells around this. Later this mass of cells acquires a skin and becomes a parasitic larva with two processes which are said to be derived from the second pair of feelers and the jaws (mandibles). Apparently the blood of the host is absorbed by these processes, which resemble the feeding processes of the Rhizocephala described in the next paragraph. Inside this larva the adult monstrillid eventually develops. When it is mature the larva (fig. 108) bores its way out of the worm upon which it has fed, throws off the feeding appendages, and bursts to set free the adult female (fig. 109), which lives a brief life on the surface of the sea, where fertilisation of the eggs occurs. The adults have, however, lost their mouth parts and the mouth is only a narrow opening, so that they cannot feed.

The suborder of Crustacea called the Rhizocephala, which are not copepods, but belong to the order Cirripedia, to which the barnacles belong, show examples of a degree of reduction and loss of organs which must represent the extreme limit to which this process can be carried. Some of them are parasitic upon either their barnacle relatives or upon starfishes and some of their relations. Among them is the remarkable genus *Sacculina* (fig. 58, see also Chapter 8), parasitic upon crabs.

The larva which hatches out of the fertilised egg of *Sacculina* is a nauplius type of larva similar to that which all the chief groups of Crustacea produce. This larva is, however, succeeded by a type of larva which is called a *Cypris* larva. This name is given to it because it is enclosed in a 'shell' composed of two halves hinged along the back, so that its shell resembles that of the non-parasitic adult ostracod crustaceans belonging to the genus *Cypris*. Because the *Cypris* type of larva is characteristic of the life history of cirripede Crustacea, the group to which the barnacles belong, we know that *Sacculina* also belongs to this group of animals. Without this evidence of its zoological relationship it would have been impossible to classify *Sacculina* correctly, because its adult phase consists of little more than a sac containing reproductive organs, which sends root-like processes into all the organs of the crab host, excepting its gills and heart.

The *Cypris* larva of *Sacculina* attaches itself to the second pair of feelers (*antennules*) of its crab host. It then throws off its limbs and becomes reduced to a mass of undifferentiated cells enclosed in a skin. By means of a small process of this mass, it penetrates into the crab's body and passes into its blood, by which it is carried to the under-side of the crab's intestine. There it forms root-like processes which penetrate into the organs of the crab's body. By means of these processes the *Sacculina* feeds upon the substance of its host. Sexual organs develop in a protuberance from it, which presses upon the integument on the under-side of the crab. The effect of this pressure is that the crab fails to form integument over the area upon which the *Sacculina* presses. The consequence is that when next the crab moults, the *Sacculina* projects as a small protuberance on the under-side of the crab's body. Its parasitic life has therefore resulted in the loss of all its organs, except the reproductive organs and feeding processes.

The examples of reduction and loss of organs among the parasitic Crustacea which have just been given are to some extent paralleled by similar examples found in the order Isopoda, to which the wood-lice belong. Species of the isopod genus *Aega*, for instance, have mouthparts modified for piercing the skin of fishes upon which *Aega* is parasitic and for sucking their blood; and their front pairs of legs have hooks for holding on to the skin of the fishes. These species can, however, like the 'fish-lice' and 'carp-lice', leave their hosts and swim about. The food canal can dilate considerably to take in a large amount of the host's blood, and in this respect they remind us of the blood-sucking leeches and bats. After a meal of blood they drop off their hosts to digest the meal at the bottom of the sea, a method of feeding which resembles that of the ticks. The blood in them, after drying and extraction, was called in Icelandic folklore 'Peter's Stone', to which magical and medicinal properties were attributed. One species of *Aega* lives in a sponge.

Much more completely committed to parasitic life are *Cymothoa* and its isopod relatives, which also cling, by means of hooks on their legs, to the skins of fishes which are their hosts. The young of these species swim about freely and their bodies are not much altered; but when they settle on a host, their bodies often lose their typical shape and become distorted. They are hermaphrodite, each individual being first male and later female.

The species of the section of the Isopoda called the Epicaridea are all parasitic upon other Crustacea. There are many of these and only two examples of them can be given. *Bopyrus squillarum*, which is parasitic in the gill cavity of the common prawn, *Leander serratus*, may cause the formation of a swelling on one side of the prawn's body in the gill region. Relatives of this species may be parasitic upon the gills of hermit crabs and their relations. Their bodies are flattened and distorted and they cling to the host's gills by means of hooks on their shortened legs. They suck the host's blood by means of mouth parts altered for this purpose. The minute male is attached to the underside of the body of the female and is parasitic upon her. Related to *Bopyrus* are the Entoniscidae (see Chapters 6 and 8), which are parasitic on the gills of crabs. Their bodies are so much modified by parasitic life that they have no external resemblance to ordinary isopod Crustacea. Other examples of the loss of organs by parasitic Crustacea, which is associated with attachment of the male to the female, are given in Chapter 8.

Many other examples could be given of the modifications or loss of organs as a result of parasitic life; but another principle may also operate. Existing organs may be modified by parasitic life in such a way that they become more useful.

Some parasitic Protozoa, for instance, which adopt life in fluid surroundings, such as blood or the contents of the food canal, which offer greater resistance to their movements than that offered by the water in which their ancestral non-parasitic life was lived, develop modifications of the number and structure of their whip-like locomotor organs (*flagella*). Thus the flagellated Protozoa which live in the intestines of white ants have developed very numerous flagella and the flagellum of the trypanosomes (fig. 18), which live in the plasma of the blood, is united to the body by a protoplasmic membrane, called the *undulating membrane*, whose undulating movements increase the locomotor function of the flagellum. The species of *Trichomonas* mentioned in Chapter 9 have also developed an undulating membrane. In *Trichomonas* and some other flagellate Protozoa a stiffening rod known as the *costa* has been developed along the base of the undulating membrane. When flagella are modified in this way, part of the nuclear material of the cell takes over the control of the flagella, so that the locomotor organs and their controlling apparatus may together form a complex system.

Among other organs which are often reduced or lost when parasitic life is adopted is the nervous system. It may be reduced as a whole or the reduction may affect chiefly the eyes and other organs. Organs of special sense are best developed in active animals which feed upon other animals and need to defend themselves against their enemies. They are not required by parasitic animals which live a relatively sheltered existence on or inside the bodies of their hosts amid a relative abundance of food. The eyes of the hagfishes, for instance, show an imperfection of structure correlated with the fact that these parasitic fishes bury their heads in the tissues of their hosts, so that the eyes are unable to function. The larval phases of some parasitic animals, however, which are not parasitic and live in the world outside the host, retain certain organs of special sense. Thus the miracidial larvae of the flukes have primitive eyes. The non-parasitic infective larvae of roundworms are sensitive to light, touch, changes of temperature and gravity. Their responses to these influences may guide them to situations in which they have the best chances of infecting their hosts. The fact that they may also guide them towards infection of abnormal or unsuitable hosts is considered in Chapter 9.

Movements of the infective larvae of some species of parasitic animals may also be influenced by chemical substances secreted by hosts into which those larvae penetrate. Thus snails, which are the intermediate hosts of some flukes, secrete chemical substances which attract miracidial larvae of these species of flukes towards their bodies. It is said that colour may also act in this way, for the sporocysts of some species of flukes which are found in the snail intermediate host are coloured and attract by their colour the birds which infect themselves with the flukes by eating the snails (see Chapter 8).

Finally, it is interesting to consider whether the reduction or loss of organs caused by parasitic life has deprived us of fossil remains which would have given us information about the history and evolution of parasitism. Because this mode of life tends to cause a loss of structure resistant enough to be preserved as fossils, we have little geological evidence of the past history of parasitic animals. At least six species of fossil roundworms have, however, been described. Two of these, *Heydonius antiquus* and *H. matutinus*, were found in the Eocene lignite and Baltic amber, while the other four species were found in Baltic amber.

Human writings about some species of parasitic animals take us back to some of our earliest records of the life of man. The Egyptian Papyrus Ebers, for example, which dates from 1600 B.C., refers to the tapeworms, blood flukes and hookworms of man. Malaria is described by ancient Chinese physicians and human hookworm disease, filariasis and Ascaris infections were known to the Persian physician Avicenna, who lived between the years A.D. 980 and 1037. The eggs of one species of human blood fluke, *Schistosoma haematobium*, have been found in the kidneys of mummies of the twentieth Egyptian dynasty, which flourished between the years 1250 and 1000 B.C. Species of human lice belonging to the genus *Pediculus* have also been found on mummies and the 'fiery serpent' mentioned by Moses in Chapter xxi, v. 6 of the Book of Numbers of the Bible was probably the guinea-worm (*Dracunculus medinensis*). The Book of Numbers also refers to plagues of fleas and lice.

CHAPTER 6

SOME EFFECTS OF PARASITIC LIFE UPON THE PARASITIC ANIMAL, II

4. MODIFICATIONS OF THE REPRODUCTIVE PROCESSES

The reader will already have realised that the parasitic animal has to contend with difficulties and risks to which its non-parasitic relatives are not exposed. It may have gained shelter and abundance of food, but it has obtained these at the cost of partial or complete dependence upon its hosts. The parasitic animal cannot live without its host. It must find it and get into its body, or on to its surface, and it must maintain itself in these situations.

When the young of the parasitic animal only are parasitic, the non‑parasitic adult can place its eggs or larvae inside the host's body or on its surface. Some species which do this are described in Chapter 9. But very often it is the young of the parasitic animal which have the task of making contact with the host and they are often vulnerable and easily become the prey of other animals. It is not surprising, therefore, that parasitic life tends to produce modifications of the reproductive organs which increase the numbers of the eggs and larvae produced. This increase of the numbers of the young compensates for the high mortality among them. How is it effected? We can sum up the chief methods by which it is accomplished by saying that:

(i) the number of eggs and sperms may be increased. When this happens, the size of the male and female reproductive glands is usually increased;

(ii) fertilisation of the egg may be made easier by a more or less permanent contact between the male and female; or by the inclusion in each parent of both male and female reproductive glands, each parent being hermaphrodite; or fertilisation may be abolished, the parasitic animal being parthenogenetic;

(iii) the number of individuals derived from each egg may be increased.

(i) *Increase of the Number of Eggs Produced*

This is well shown by some species of parasitic worms. The increase of the number of eggs produced may be effected by an increase of the size or the number of the ovaries which produce them and perhaps also of the testes which produce the sperms; or by an increase of the capacity of a relatively small ovary to produce eggs.

The former method is well shown by many species of flukes and tapeworms. The hermaphrodite reproductive organs of the flukes, together with the glands which produce yolk and shells for the eggs, occupy, when they are active, the greater part of the body of the fluke. These organisms have, however, only one set of them in each individual.

The hermaphrodite reproductive organs of the tapeworms, on the other hand, not only have a great capacity for producing eggs, but are serially repeated in each of the segments of which the whole tapeworm is composed. There is a complete set of male and female organs and their accessory glands in each segment of the tapeworm, so that the egg-production of the tapeworm individual is greatly increased. The segments are not all sexually mature at the same time, the young sexually immature ones being nearest the head, while the rest of the chain consists of the older ones in various stages of sexual development until those at the free end of the worm are producing fertilised eggs. Each segment of some species of tapeworm can fertilise its own eggs, but the eggs of each segment of other species must be cross-fertilised by sperms from other segments.

The numbers of eggs produced by some species of tapeworm are so great that it is difficult to realise how enormous they are. When the eggs have been fertilised, each segment of the tapeworm contains little else but the distended uterus crammed with eggs.

It has been calculated that the fish tapeworm of man, *Diphyllobothrium latum*, produces about 36,000 eggs a day and may produce a million daily; that the beef tapeworm of man, *Taenia saginata*, produces between 50 and 150 millions of eggs a year and that each of its segments, of which there may be 1000–2000, may expel an average of 97,000 eggs. All estimates of this kind should take into account the length of life of the parasitic animal concerned (see below). *T. saginata*, for instance, can live in man for 10 years, so that it can during this time produce a remarkable number of eggs.

The pork tapeworm, *T. solium*, is less prolific. It has been calculated that each of its segments, of which there may be 800–900, can produce 850,000 eggs or more, so that its total egg-production would be about 700 or 800 millions.

The production of very large numbers of fertilised eggs is also well shown by some, though not all, species of parasitic roundworms. The male and female reproductive organs of roundworms are in separate individuals and have the form of long coiled tubes which can be accommodated in the

cylindrical bodies of these worms. They are often volu-
minous, but the egg-production of some species of roundworms is much greater than the size of their ovaries would
suggest.

It has been estimated, for instance, that the female large
roundworm of man and the pig, *Ascaris lumbricoides*, may
contain, at any one time, 27 million eggs; and that each
female of this species can lay 200,000 eggs a day. Other
experts place the egg-production of this species at 60 million
eggs per year, some 75 % of which may remain able to
infect new hosts for many months. Probably this enormous
egg-production is associated with the fact that *A. lumbricoides* lives only for a year or less. It has been estimated that,
if each pair of ascarids produces 200,000 eggs a day, the
annual egg-production of each pair weigh about 5 grams.
In China, where some 335 million people were found by
American workers to be infected with this species, the weight
of *Ascaris* eggs produced each year 'in the whole infected
Chinese population' would, on this computation, be
18,000 tons.

Estimates of the egg-production of *Ancylostoma duodenale*,
one of the human hookworms, state that this species
produces from 25,000 to 35,000 eggs per day, and that
only 5 % of these are present in the uterus of the worm at
any one time. This species of hookworm may live for
5 years or so, so that it may lay, during this period, some
18–54 million eggs. It has been estimated, however, that
the chances against a male and a female hookworm getting
into the same human host and living successfully in it are
about 18 million to one, so that every female will not
succeed in laying this number of fertilised eggs. The peak
of egg-production is, moreover, reached in six months or

so and after this time many hookworms die or their egg-production falls. It is fortunate for man that these limitations exist, because if we take the lowest of the estimates of the daily egg-production given above and consider it in relation to the estimate of American experts that 500 hookworms must suck blood for several months in order to cause severe hookworm disease we must conclude, if we make the reasonable assumption that half of these 500 hookworms were females, that the minimum number of eggs produced every day in each person suffering from this degree of infection would be $250 \times 25{,}000$, or about 6 million eggs.

The other species of hookworm which infects man, *Necator americanus*, produces fewer eggs. It has been calculated that its females produce 6000 to 20,000 eggs per day. Estimates of its length of life vary from 5 to 17 years. The more moderate egg-production of the dog hookworm, *Ancylostoma caninum*, is mentioned below.

Some species of roundworms which live in the food canals of sheep, cattle, horses and other domesticated animals also produce large numbers of eggs. It has been estimated, for instance, that pastures grazed by a group of apparently healthy sheep infected with these roundworms may receive from these sheep 8 million roundworm eggs daily and that horses infected with related species of roundworms may pass out a similar or larger number of the eggs of these worms.

There are, nevertheless, species of roundworm that live in the food canals of sheep which produce much smaller numbers of eggs. *Cooperia curticei*, for instance, produces only 400–900 in 24 hours. Some species of roundworms which lay, not eggs, but living larvae are, on the other hand, just as prolific as the species just mentioned. The female guinea-worm, *Dracunculus medinensis*, for example, may liberate millions of larvae from her uterus.

The egg-production of all species of parasitic roundworms, and probably that of other kinds of parasitic animals also, is not, however, constant. It varies from time to time and is influenced by the genetic constitution of the parasitic animal, the resistance of the host to it and other factors, some of which are not yet fully understood. The number of roundworms present in the host may influence it, for fewer eggs are produced when a larger number of adult roundworms is present. This may be the result of overcrowding and competition for nourishment, and it is possible that premunition, described in Chapter 7, also plays a part.

That the host undoubtedly does influence the number of eggs produced is clearly shown by experiments which have been done with the physiological strains (see Chapter 7) of the dog hookworm, *Ancylostoma caninum*, which are found in the dog and the cat respectively. When the strain of this species found in dogs was put into other dogs, each female of it produced some 16,000 eggs per day; but when it was put into cats, each female of it produced only 2340 eggs per day. When the strain found in the cat was put into other cats, each female of it produced 2340 eggs per day; but when this strain was put into dogs, each female of it produced as many as 11,600 eggs per day. The females of both strains thus produced more eggs in dogs than in cats. These experiments show that the rate and degree of egg-production of roundworms, at any rate, is not inherent in the roundworm, but is influenced by other factors whose nature we do not yet understand.

The figures just given will be sufficient to indicate that some species of roundworms produce enormous numbers of eggs. So great, indeed, may these numbers be that the shape of the female roundworm may be altered by the mass of eggs which she is carrying. In species of the genus *Tropisurus* (*Tetrameres*), for instance, a roundworm which causes disease of the part of the food canal (*proventriculus*) of ducks, pigeons, turkeys, poultry and other birds which precedes the gizzard, the females become distended by their eggs and lose the cylindrical shape characteristic of round-worms. They become fusiform or almost globular as they lie partly embedded in the glands which secrete the gastric juice, while the males, which live in the contents of this part of the food canal, are slender and cylindrical. The thorny-headed worm, *Macracanthorhynchus hirudinaceus* (cf. Chapter 5), which is usually classified as a relative of the roundworms, produces very hardy eggs, enclosed in four membranes which can live for several years and may lay 260,000 of them every day for a period of 10 months or so.

The number of eggs produced by parasitic insects and arachnids is smaller. The human head-louse produces 80–100 eggs and the human body-louse 200–300, but the human pubic (crab-) louse produces only 40. The female warble-fly, *Hypoderma lineatum*, lays only about 100 eggs on each cow, and each female may lay 500–800 eggs. The female horse bot-fly, *Gastrophilus intestinalis*, has been watched while she was laying eggs on the hairs of a horse and she laid 301 eggs in 45 minutes. Inside other females of this species from 700 to more than 1000 eggs were found. The sheep ked, *Melophagus ovinus* (fig. 88), lays, not eggs, but living larvae and produces only about 10-15 of these, at the rate of about one each week, during her life-time of about 16–20 weeks.

Among the Arachnida the ticks produce more eggs than the species just mentioned. The castor-bean tick, *Ixodes ricinus*, which feeds only once in her lifetime, lays 500–2000 eggs in crevices in the ground after a feed (fig. 12). This is a three-host tick (see Chapter 9), that is to say, a species whose larvae, nymphs and adults feed upon different individual hosts. The whole life history, moreover, occupies, in Britain, three years, the larva, nymph and adult each feeding once each year. There is therefore plenty of risk that the life history may not be completed and a large number of eggs compensates for this risk. One-host ticks, such as *Boophilus annulatus*, are exposed to less risk, because their larvae, nymphs and adults all feed upon the same individual host. They produce, on the whole, fewer eggs than the three-host ticks do. The soft ticks, such as *Argas persicus*, lay eggs after each feed; they take a feed about once a month, usually at night.

It is tempting to suppose that species such as the warble-flies, which attach their eggs to the hairs of their hosts, lay fewer eggs than some other parasitic insects because the adult places the larva in a position favourable to its successful penetration into the host.

Although this may be one reason why fewer eggs are produced by some species of parasitic animals, we should not forget that egg-production is influenced also by the factors, already mentioned, such as the genetic constitution of the mother, her state of nourishment and so on, all of which operate in the parent rather than on the adaptational capacity which affects the life history as a whole.

The mere production of large numbers of eggs is not

sufficient by itself to ensure the survival of the species. Both the eggs and young which hatch out of them must be able to live in the environment into which they pass. Eggs and young phases therefore show adaptations which help them to resist the injurious influences in these environments.

These adaptations are especially well shown, perhaps, by species of parasitic animals whose eggs and young must live for a time a non-parasitic life and must therefore be able to survive entry into the external world. For this reason they are enclosed inside the resistant envelopes which surround the cysts and spores of Protozoa and the eggs of parasitic worms. Frequently the resistance of these structures to climatic conditions depends largely upon some form of fatty envelope inside an external membrane composed of chitin or some other resistant material. This use of fatty material for protection may be compared with the protection which waxy layers afford to insects and to some species of bacilli, such as the species which causes tuberculosis.

The young of some species of parasitic single-celled Protozoa may be provided with extensions of their body surfaces and other devices which prevent them from sinking in the sea or fresh water in which they are deposited; they are thus prevented from sinking down too far away from the aquatic hosts which they infect. This feature is shown by the spores of Protozoa called Myxosporidia and Actinomyxidia, which infect fishes.

(ii) *Modifications which Help the Fertilization of the Eggs*

Among the simplest of these are modifications which help to maintain the union of the male and female. An example of them is the *copulatory bursa* (fig. 110) characteristic of male roundworms belonging to the order Strongyloidea.

This is an umbrella-like expansion of the cuticle of the hind end of the male, supported by rays of stiffer material, which extends over the male genital openings. It is applied over the female genital openings and maintains a hold

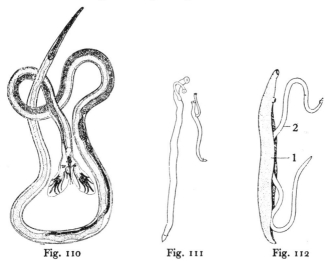

Fig. 110   Fig. 111   Fig. 112

Fig. 110. The large stomachworm of sheep, *Haemonchus contortus*, showing the fan-like copulatory bursa at the hind end. Female 18–30 mm., male 10–20 mm.

Fig. 111. The 'gapes' worm of poultry, *Syngamus trachea*, showing the male and female with their genital openings permanently applied to each other. Female 5–20 mm., male 2–6 mm. long.

Fig. 112. Male and female blood flukes of the genus *Schistosoma*, showing how the male forms a groove for the accommodation of the more thread-like female (diagrammatic). *S. haematobium*, female 20 mm., male 10–15 mm. long; *S. mansoni*, female 7–14 mm., male 6·5–10 mm. long; *S. japonicum*, female 26 mm., male 12–20 mm. long.

there. Its efficiency may be reinforced by hooks or papillae on these regions of the roundworm's body. The smaller male of the 'gapes' worm, *Syngamus trachea* (fig. 111) is more or less permanently united in this way to the larger female, and the Y-shaped pair thus formed no doubt resists better their separation by the movements of air and frothy mucus in the windpipes of birds in which they live.

A unique example of permanent union of the sexes is provided by *Diplozoon paradoxum*, one of the monogenetic flukes. The hermaphrodite adults of this species are parasitic on the gills of freshwater fishes (carp, stickleback, etc.). The larvae of this species are single individuals parasitic on the gills of the minnow and other cyprinid fish, but they become permanently united in pairs by union of their body tissue in the region of the genital openings, the sexual ducts of one individual being continuous with those of the other.

Another kind of adaptation which effects the same result is the folding of the hinder four-fifths of the body of the male flukes belonging to the genus *Schistosoma* (fig. 112) to form a groove in which the female is held. The body of the female has lost the flattening characteristic of the flukes and has become cylindrical to fit the groove. No doubt these changes are correlated with the life of these species in the narrow channels of the lymphatic system of man, where they are exposed to the slow but continuous currents of the lymph.

This process of temporary or permanent union of the sexes is carried even further by some species of roundworms whose males are parasitic upon the females. The diminutive male of the species *Trichosomoides crassicauda*, parasitic in the uterus of the female of the same species is mentioned in Chapter 9. Among non-parasitic animals there are three species of abyssal fishes, *Photocorynus spiniceps*, *Ceratius holbelli*, and *Edriolychnus schmidti*, whose dwarf males are parasitic upon the exterior of the females.

Attachment of the male to the body of the female,

combined with a difference of size or of other characters between the sexes (*sexual dimorphism*), is also found among the parasitic copepods mentioned in Chapter 5. The males of some species of the Chondracanthidae and Lernaeopodidae are attached to the bodies of the females; and the males of some blood-sucking species belonging to the family Choniostomatidae, which are parasitic upon other Crustacea (Malacostraca), are small and are attached to the body of the female by means of a stalk. Some species of the isopod family Bopyridae (see also Chapter 5) have dwarf males, and the related family, the Entoniscidae (see also Chapters 5 and 8), show two kinds of very minute males which are enclosed with the deformed females inside the envelope formed by the infoldings of the gill chambers of their crustacean hosts.

Related to these two families is the hermaphrodite family Cryptoniscidae, species of which are parasitic upon the Rhizocephala described in Chapters 5 and 8 and upon other cirripede Crustacea, and the family Cyproniscidae, parasitic upon ostracod Crustacea. The adult cryptoniscid loses all or most of its limbs. The males of the genus *Cyproniscus* are attached to the females, which become little more than sacs of eggs. This is an example of reduction of the female to little more than a sac which incubates the eggs, a condition which is correlated with the attachment of the male to the female and sometimes also with hermaphroditism.

The crustacean order Cirripedia, to which the barnacles belong, includes species whose minute males, which are called *complemental males*, are attached to larger individuals which are either females or are hermaphrodite, or the relationships of these minute males to the females or hermaphrodite individuals may take various other forms.

While these modifications which aid the fertilisation of the eggs are interesting, they affect the actual reproductive processes of the parasitic animal less than certain other modifications which assist the fertilisation of the eggs do. Two of these modifications will be discussed briefly here, namely, the abolition of the necessity for the fertilisation of the egg, which is called *parthenogenesis*, and the inclusion of both sexes in the same individual, which is called *hermaphroditism*.

It would be difficult to maintain that either parthenogenesis or hermaphroditism is a direct result of the adoption of the parasitic life. Parthenogenesis is, in fact, not at all common among parasitic animals and is practically confined to one or two species of ticks, such as *Amblyomma rotundatum*, and to some parasitic insects belonging to the group to which the ants, bees and wasps belong (Hymenoptera). It is, indeed, much more characteristic of non-parasitic than of parasitic animals, a familiar example being the honey bee, males of which are developed from unfertilised eggs laid either by the queens or by fertile workers. Other examples are shown by other Hymenoptera. Parasitic Hymenoptera which practise it often combine it with ordinary sexual reproduction and produce only males by parthenogenesis.

Hermaphroditism is also characteristic of many non-parasitic animals. All the ringed, bristle-bearing worms (Annelida), for example, to which the earthworms and leeches (figs. 5, 6) belong, are hermaphrodite, and sedentary animals show a tendency to adopt this arrangement of their reproductive organs, perhaps because their adult phases, like those of some parasitic animals, find it relatively difficult to ensure that their sperms and eggs are produced as near to each other as possible.

Among parasitic animals hermaphroditism is characteristic of the flukes and tapeworms, and the fact that it
is the reproductive method of these two large groups of
parasitic animals may seem, at first sight, to suggest that it
has resulted directly from their parasitic life. It seems just
as likely, however, that it is not a result of the parasitic life
of flukes and tapeworms, but a feature inherited from their
non-parasitic ancestors. What these ancestors were we do
not know, but it could be argued that a parasitic animal
which carried with it from the very beginning of its parasitic life the union of the two sexes in the single individual
would begin parasitic life with a definite advantage; and
the success of the flukes and tapeworms as parasites may
have been partly due to the fact that they began their
parasitic life with this advantage in hand. Their hermaphroditism may, on the other hand, be due to the same
causes which produce hermaphroditism in sluggish or
sedentary animals. Whatever the reason for their hermaphroditism may be, it must absolve them from the difficulties that beset the fertilisation of the eggs of unisexual
animals.

(iii) *Increase of the number of individuals derived
from each fertilised egg*

The result of the two kinds of modification of the reproductive processes just described is an increase of the total
number of potentially infective phases, whether these be
eggs or larvae, that are derived from the eggs. The number
of infective phases, however, each of which is potentially
an adult parasitic animal, may be increased by another
method. The larva or other phase which leaves the egg or
its equivalent may itself multiply its numbers before the

infective phase is reached, so that each egg or its equivalent gives rise to a large number of potentially parasitic
adults. This method of increasing the parasitic animal's
chances of survival is shown by the digenetic flukes and
by some species of tapeworms and parasitic Protozoa.

Among the Protozoa it is effected by the asexual process
known as schizogony. The multiplication effected by the
flukes and tapeworms may be either asexual or an extension of the sexual process into the larval phases. Yet other
flukes use a method which appears to be the sexual process
called *polyembryony*, which is the production of a number of
embryos from a single fertilised egg. Polyembryony is
shown also by some parasitic insects belonging to the
order Hymenoptera. The monogenetic flukes mentioned in
Chapter 5 omit this multiplication of larval forms, but
some of the larvae of one species of them, *Polystoma
integerrimum*, whose life history is described below, appear
to be able to produce eggs (*neotony*).

The tapeworms that multiply their number of potential
adults of the species produce more than one tapeworm
head on the walls of the bladderworms. When the germinal
membrane lining the bladderworm produces only one
tapeworm head, that bladderworm is called a *cysticercus*
(fig. 48); when it produces many heads, the bladderworm
is called a *coenurus* (fig. 49). In the hydatid cysts (fig. 50)
the germinal membrane does not directly produce tapeworm heads, but produces stalked bladders, called *brood
capsules*, in each of which as many as forty tapeworm heads
may arise.

These brood capsules may remain attached to the wall of
the hydatid cyst or they may be set free, together with the
tapeworm heads inside them, into the cavity of the mother

cyst; and some tapeworm heads may also be set free into this cavity. The free heads and brood capsules together form a deposit in the cyst which is called *hydatid sand*. If the hydatid cyst is injured and ruptured and portions of it are carried elsewhere in the host's body by the blood, the germinal membrane of these detached portions may produce, in the new situation taken up, daughter and even granddaughter cysts, each of which can be as prolific as the mother-cyst from which it originated. Although some hydatid cysts fail to produce any tapeworm heads at all, multiplication in the fertile ones may go on for as long as 10 or 20 years, and it has been estimated that as many as two million tapeworm heads may arise inside a single hydatid cyst. All these have originated from a single fertilised egg and all of them are structurally alike.

The facts that a single dog in which the adult *Echinococcus granulosus* produces eggs may harbour hundreds or even thousands of adult tapeworms of this species and that each ripe segment of this tapeworm may produce 500–800 fertilised eggs, give us an idea of the capacity of this species to multiply its individuals. It is fortunate that this species of tapeworm consists of only three or four segments, only one of which is ripe and ready to discharge fertilised eggs at any one time. Its egg-production, moreover, is much smaller than that of the species of tapeworms mentioned earlier in this chapter, and many hydatid cysts formed from its eggs fail in some definitive hosts to produce any tapeworm heads at all. In cattle, for instance, only about 10 % of them are fertile, although in pigs about 80 % and in sheep about 92 % contain living tapeworm heads capable of infecting dogs. Against these limitations must be set the power of the germinal membrane to produce, if the cyst is injured or ruptures, the daughter and even grand-daughter cysts mentioned above.

The flukes which multiply their larval phases produce, in contrast to the tapeworms, not larval forms which are structurally all alike, but the sporocysts, rediae and cercariae (figs. 40–45) described in Chapter 4, which are very different from each other.

When a single miracidium larva (fig. 41) leaves the egg of the liver fluke of sheep and cattle, *Fasciola hepatica*, and enters the snail which is the intermediate host of this species, it changes into a sporocyst (fig. 42), which produces a number of rediae (fig. 43), each of which may give rise to more rediae. From the rediae arise cercariae (fig. 44), which become the metacercariae (fig. 82) that alone can infect the definitive host.

Different species of flukes increase the number of larval phases at different points in the sequence of these larval forms. Thus some species produce two or more generations of rediae; others, such as *Schistosoma mansoni*, one of the blood flukes of man, produce no rediae, but rely upon a second generation of sporocysts, which goes on producing cercariae for several days or weeks. It has been estimated that the single miracidium larva derived from each egg of this species may give origin, inside a single snail intermediate host, to an average of 3500 cercariae a day for a considerable time, and that the total number of cercariae thus derived, during a period of months, from a single miracidium larva may be 100,000–250,000. Because each of these cercariae, if it enters man or another suitable definitive host, may grow into an adult blood fluke, the potential degree of the infection of man from a single snail is considerable.

It is fortunate for man that this relatively enormous multiplication of infective individuals derived from each egg is offset by the fact that the miracidium larva which leaves the egg to give rise to them can live only about 16 hours, during which time it must find the species of snail in whose body it can live, or it must perish. It is fortunate also that the entry of a single cercaria of a blood fluke into man cannot cause the serious disease (schistosomiasis) which these flukes produce. These flukes are, like the round-worms, unisexual, so that they can complete their life histories only when enough cercariae enter man to secure the survival in him of both male and female flukes. Only then can the flukes produce the fertilised eggs, the exit of which is the cause of much of the harm done by these flukes.

The multiplication of larval phases in this way has been given as a reason why the flukes as a whole produce fewer eggs than the roundworms, each egg of which can give rise to a single adult only. Unfortunately for this argument, some species of flukes produce large numbers of eggs. Thus the Asiatic giant intestinal fluke of man and pigs, *Fasciolopsis buski*, produces some 25,000 eggs a day and, in addition, multiplies its larval phases by producing two generations of rediae, its life history resembling that of *Fasciola hepatica*.

Among the parasitic Protozoa the numbers of infective phases may be increased by division of the zygote itself to form sporozoites. Division of the zygote of *Eimeria* produces only eight individual sporozoites, but division of the zygote of the malarial parasites inside the soft-walled cyst in the wall of the mosquito's stomach gives rise to some 10,000 sporozoites, and the salivary glands of an infected mosquito may contain as many as 60,000 sporozoites at one time. Each of the sporozoites thus produced is potentially able to grow up, in suitable surroundings, into a trophozoite.

The multiplication of individuals thus effected may then be supplemented by multiplication of individuals already parasitic. This is effected by the asexual process of schizogony. In *Eimeria* this schizogony occurs in the host in which the division of the zygote occurred. The schizogony of the malarial parasites, however, occurs in the intermediate host. At each schizogony in both species the number of individuals produced is relatively small. *E. caviae* produces at each schizogony 12–32 merozoites, *Plasmodium vivax* produces 12–24, *P. falciparum* 24 and *P. malariae* 8. But the repetition of schizogony may produce enormous numbers of descendants.

Thus *Eimeria tenella* can kill young chickens in 4 or 5 days because 225,000 merozoites of the second generation can arise from each of its trophozoites, so that infections with a single oocyst, which contains 8 sporozoites, each of which can give rise to a trophozoite, may result in the formation, in 4 or 5 days, of 1,800,000 merozoites, each of which may destroy a cell of the chicken's caecum. This rapid multiplication also causes profuse bleeding, to which the deaths of the chickens are largely due.

Multiplication of malarial parasites in the liver and blood by schizogony may result in the blood being crammed with the parasites. The blood of children in some parts of Africa during the malaria season may contain several times as many malarial parasites as red blood cells. It has been calculated that *P. vivax* can produce about 40,000 individuals in every cubic millimetre of the host's blood. *P. malariae* is less prolific, but *P. falciparum* can produce, according to some estimates, a total of some 3 billions of individuals in the blood of a single human being, a quantity whose volume would fill a pint pot.

5. CORRELATIONS OF THE LIFE HISTORY OF THE PARASITIC ANIMAL WITH THAT OF THE HOST

The fact that parasitic life produces profound changes in the life history must be already evident to the reader. Less evident will be the interesting correlations that exist between the histories of parasitic animals and those of their hosts. A good example of them is provided by the mono-genetic fluke, *Polystoma integerrimum* (fig. 75), which lives in the bladder of the common frog, where it can be easily seen through the transparent walls of this organ. It is attached to the walls of the bladder by six suckers arranged in a crescent.

This species of fluke lays its eggs in water and it lays them in the spring, when the frog visits water to lay and fertilise its own eggs. During the winter the reproductive organs of the fluke are dormant. In the spring, however, it produces about 1000 eggs, laying about 100 a day into the water. The eggs hatch either at the end of April or during May and June.

The miracidium which hatches out of each egg swims about in the water by means of its circlet of cilia, and its behaviour is aimless until it meets with a tadpole which is old enough to have lost its first finger-like external gills and to have developed, instead of these, the internal gills situated on the walls of gill slits at the sides of the tadpole's throat. These internal gills mark the fish-like phase of the frog's life history. They are covered in by two folds of skin which grow back to form a bag called the *operculum* around them, and this bag opens to the water by means of a spout-like opening on the left-hand side of the tadpole. Water taken in through the tadpole's mouth passes into its throat, out through the gill slits and over the internal gills in them and so into the bag around them and out by the spout-like opening.

The miracidium of *P. integerrimum* ignores all tadpoles which have not yet reached this stage of their development; but when it meets with one which has, its aimless behaviour ceases. It seems to pause and to await its opportunity to dart through the spout-like opening into the bag around the internal gills. How it knows that the tadpole has reached this stage of its development we do not know, but perhaps it is helped by its eye-spots and its nervous system or by chemical substances secreted by the tadpole into the water which stimulate the miracidium larva.

Once it is inside the gill chamber the miracidium is attracted to the internal gills and attaches itself to these by the sixteen hooklets on a sucker-like structure at its hind end. There it feeds for 8–10 weeks on the mucus and debris which cover the gills. During this period a great many *Polystoma* larvae are lost from the gill-chambers of the tadpoles, and many die because many of their tadpole hosts die. The *Polystoma* larvae therefore have the best chance of becoming adults if they enter the tadpoles near the time when the tadpole becomes a frog. In ponds in which this change is delayed until July or August some of the *Polystoma* larvae produced earlier may produce eggs which provide miracidia in time to infect these tadpoles which develop late.

When the gills of the tadpole disappear and the tadpole

becomes an air-breathing frog, the fluke larva loses its cilia and develops the six powerful suckers of the adult fluke described in this chapter. It also leaves the gill-chamber and passes into the gullet of the frog and down its food canal. When the frog's bladder develops, the young fluke enters this to attach itself by means of its six suckers to the wall of the bladder and to take up its adult life there. It does not become sexually mature until it is 3 years old.

It is said, however, that if the young fluke attaches itself to tissues which are full of blood, it may become sexually mature in 5 weeks. The delicate external gills of the young frog before it begins to breathe like a fish by means of internal gills are, of course, richly supplied with blood, and if the miracidium larva attaches itself to these by mistake, it develops so quickly that it becomes an adult and produces eggs before the tadpole becomes an air-breathing frog. When this happens, it never reaches the frog's bladder, but dies when the young frog begins to breathe air. This occasionally happens even when the miracidium attaches itself to the internal gills of the tadpole in the usual way.

This life history illustrates very well the manner in which the life histories of the parasitic animal and the host may be intimately correlated. The eggs of the fluke are laid at a time when the miracidium larvae have the best chance of becoming parasitic on the internal gills of the tadpoles, and the change from the larval phase of the fluke to its adult phase coincides with the change of the water-breathing tadpole to the air-breathing frog, which removes from the larval fluke its food supply provided by the tadpole's gills.

The timing of the release of the parasitic animal's reproductive phases so that they may infect the host required is also shown by some species of Protozoa which live in the rectum of the frog. These species, which belong to the genera *Balantidium, Opalina* and *Nyctotherus*, produce, not eggs, but cysts containing phases which propagate the species in the new host, and these cysts are produced only in the spring, when the tadpoles, which eat them and so infect themselves, are also produced.

Many other instances of similar correlations between the life histories of parasitic animals and their hosts could be given. Some of them have been noted already in the course of this book. We have already mentioned, for example, the appearance in human blood of the microfilarial larvae of Bancroft's filarial roundworm only during the night, when the mosquitoes which are the intermediate hosts of this species bite and suck blood. We have also mentioned the corresponding diurnal periodicity of the appearance in human blood of the microfilarial larvae of the human eye-worm, *Loa loa*, whose intermediate hosts bite man during the day. Another example mentioned in Chapter 9 is the arrest of the development of the larvae of the hook-worms of dogs and of ascarids of cattle while these are in the unborn embryos of these hosts.

Among species which are parasitic in insects there are many examples of correlations between the life histories of host and parasite which enable the parasitic animal to survive during the changes which occur when the insect larva becomes a pupa and the pupa produces the adult insect. Some of these are indicated in Chapter 9 and another example of them is provided by the parasitic larvae of the roundworm genus *Habronema*. Three species of this genus are parasitic in the stomachs of horses and their relatives. Their larvae are passed out in the droppings of the horses and are eaten by the maggots of house-flies or stable-flies, in which they are parasitic. They survive through the pupation of the maggot to become, at the time when the adult fly is formed, infective larvae capable of infecting the horse again. The infective larvae make their way into the proboscis of the fly. The larvae of *H. microstoma*, which are parasitic in the blood-sucking stable-fly, *Stomoxys calcitrans*, are not injected into the horse when this fly sucks blood, but break out of the fly's proboscis, much as the infective larvae of filarial roundworms break out of the proboscis of mosquitoes, and then enter the horse through the wound made by the bite of the fly. The larvae of *Habronema megastoma* and *H. muscae* are parasitic in the house-fly, *Musca domestica*, which does not suck blood. These larvae also may break out of the fly's proboscis when the fly feeds on the lips of the horses or on wounds on its surface; but more usually they are swallowed when the horse swallows the whole fly. *Habronema* larvae which get into wounds on the skin of the horse may irritate these and prevent their healing, so that a form of inflammation of the skin called *summer sores* may develop.

Other parasitic roundworms show similar correlations of the life histories with those of their insect hosts. Thus the eggs of some relatives of the rainworm are fertilised before the pupa of the fly in which they are parasitic is formed, so that these roundworms can pass on into the adult flies. This device is also shown by some parasitic Hymenoptera. Interesting also is the arrest of the development of the two-winged fly *Lucilia bufonivora*, while its host, the common toad, *Bufo vulgaris*, in whose nasal cavities the larvae of this fly are parasitic, hibernates through the winter. The larvae of the hairworm, *Parachordodes (Gordius) tolosanus*, pass the winter in a similar way inside the aquatic larvae of the alder-fly, *Sialis lutaria*, into whose bodies the hairworm larvae, set free from eggs laid in the water, bore by means of spines on their heads. They may also bore into the larvae of mayflies or those of other insects which have aquatic larvae. In the muscles or fat bodies of these aquatic larvae this hairworm passes the winter, and in the spring it passes on into the adult fly. If, in the spring, the adult insect is eaten by a ground-beetle (*Pterostichus*) or other predatory insect, the hairworm develops into the adult worm. This adult eats up all the fat body and perhaps also the other internal organs of the beetle or other host and then bores its way out of this host to mature in water or damp soil. These worms may be 5 inches long and occasionally they are swallowed by man, but it is doubtful whether they can do him any harm.

The larvae of the hairworms may also develop in fresh-water worms or fish.

THE DURATION OF THE LIVES OF PARASITIC ANIMALS

Estimates of the duration of the lives of parasitic animals may be of three kinds. Either we can attempt to estimate the duration of the whole life history considered as a biological unit or we can restrict ourselves to estimates of the duration of particular phases of each life history or to the duration of the life of the adult parasitic animal.

Estimates of the duration of the whole life history are complicated by the fact that, whether the parasitic animal uses an intermediate host or does not, it may multiply, by the methods described above, the number of individuals derived from each egg and we cannot estimate accurately the duration of the lives of all these individuals. We can, however, estimate with greater accuracy how long the whole period of this multiplication of the number of individuals usually lasts.

We know, for example, that, provided no new infection occurs, *Plasmodium vivax*, the cause of benign tertian malaria of man, may persist in man for as long as 10 years, although it usually disappears from him within 2 years or so; we know that *P. malariae*, the cause of quartan malaria, may persist in man for 3–5 years; and that *P. falciparum*, the cause of malignant tertian malaria of man, persists only for 6–12 months or so. Estimates like this of the duration of particular phases of each life history provide us with information which is essential if we are to understand the spread of diseases caused by parasitic animals, and they may provide us with a basis for the control of the parasitic animals concerned.

The importance, for instance, of knowing the time required by the non-parasitic larvae of such a species as *Haemonchus contortus* to complete their development outside the host has been mentioned in Chapter 3. It tells us when this species can become infective and enables us to take practical measures to prevent the infection of sheep. For a similar reason it is valuable to know how long the infective larvae of human hookworms must live and develop in the soil before they can infect man, and how long the malarial parasites and the filarial roundworms require to complete their development to the infective stage inside the bodies of the mosquitoes which transmit them to man. The period of time during which trypanosomes which cause disease may persist in tsetse flies which transmit them is for the same reason important.

Estimates of the duration of the adult phases of parasitic animals are no less important for the control of diseases caused by them. It is, for example, valuable to know, as we do, that the adult phases of the human hookworms may live in the human intestine as long as 16 years, although most of them probably die within as many months.

When we attempt to make out estimates of the duration of particular phases of the life history, we have to remember that the life of the parasitic animal is menaced by factors which operate upon the non-parasitic phases on the one hand, and upon the parasitic phases on the other; and that the duration of the life of the parasitic animal inside its host is determined, as its egg-production also is, not only by its genetic constitution, although this is an important factor, but also by the resistance of the host to it. If it is living in its usual host, it will, as a general rule, live longer than it will in an unusual host which reacts more vigorously against it. Any of the factors discussed in Chapter 7, which diminish the resistance of the host, may help the parasitic animal to live longer.

It will be clear, therefore, that any estimates of the duration of the lives of adult parasitic animals or of their endoparasitic larval stages must be approximate only. Some of these estimates are given elsewhere in this book. To these may be added here the following brief notes on the longevity of the adults of species which are parasitic in man and his domesticated animals.

Nematodes (Roundworms)

Haemonchus contortus, parasitic in the abomasum (stomach) of sheep, cattle and other ruminants	5 months or less but, in some hosts, eggs may be laid for at least 14 months
Ascaris lumbricoides, parasitic in the small intestine of the pig (and man)	1 year or less (in man)
Trichinella spiralis, parasitic in the small intestine of man, the pig, the rat, etc. The larvae of this species may remain alive in the muscles of its host and able to infect other hosts for as long as 31 years	2 to 5 weeks
Enterobius vermicularis, the human threadworm	3 weeks. The female then lays her eggs and dies
Ancylostoma duodenale, one of the human hookworms	A few months to 6 or 7 years, or as long as 16 years according to some records
Necator americanus, one of the human hookworms	15 years, but probably much less. Some records suggest 5–17 years
Loa loa, the human eye-worm	4 to 17 years
Acanthocheilonema perstans, the cause of one form of human filariasis	3 to 4 years
Dirofilaria immitis, the heart-worm of the dog	Several years

Trematodes (Flukes)

Species of *Schistosoma*, the cause of human schistosomiasis	29 to 30 years, according to some records

Cestodes (tapeworms)

Taenia solium, the pork tapeworm	Up to 25 years or longer
Taenia saginata	10 years

Arthropods

Males of *Sarcoptes scabiei*, the cause of human scabies	4 to 5 weeks
Psoroptes communis var. *ovis*, the cause of sheep-scab	
On the host, male	Up to 34 days
female	Up to 42 days
Off the host, male and female	10–17 days
The louse	30 days or so
The human flea, *Pulex irritans*	19 months, but variable according to the temperature, humidity and food available
The warble-fly	5 or 6 days

SOME EFFECTS OF PARASITIC ANIMALS UPON THEIR HOSTS, TISSUE REACTIONS AND RESISTANCE (IMMUNITY) OF THE HOST

Different kinds of parasitic animals affect their various hosts in a number of different ways, and it is not easy to give, in a short space, an accurate account of the different kinds of effects that they may produce. Broadly speaking, these effects can be grouped into two main categories. These are:

(1) Changes in the host which can be called its *defence reactions* against the parasitic invader. These reactions express the host's effort to localise and overcome the injurious effects of the parasitic animal, to repair any damage done and to kill and remove the parasitic animal itself. Defence reactions can, for descriptive purposes, be divided into: (*a*) *Tissue reactions* and (*b*) *Resistance* (or *Immunity*).

Tissue reactions are changes in those tissues of the host which are directly affected by the parasitic animal. They are therefore usually localised, being seen in the immediate neighbourhood of the parasitic animal or in tissues through which it has passed.

Resistance (*Immunity*), unlike the tissue reactions, is a response of the host acting as an organised whole organism. Two important consequences follow from the fact that it is a response of the whole organism. First, the response calls into action tissues and organs of the host situated at a distance from the parasitic animal. Secondly, resistance usually operates, not only at the point affected by the invader, but also all over the body of the organism which develops it. There are, it is true, some instances of what seems to be a purely localised resistance (immunity), but usually the resistance reaction is generalised.

Tissue reactions and resistance, although they have been thus separated for descriptive purposes, are, in the living host, closely interconnected. Neither can begin until the parasitic animal has made contact with the tissues or tissue fluids of the host, but both may begin immediately after it has done this. Resistance usually develops more slowly to its maximum than the tissue reaction does; and it may last a much longer time, sometimes, indeed, persisting throughout the lifetime of the host.

(2) The second category of effects upon the host includes those which cannot be conveniently described as either tissue reactions or resistance, although both or either of these may result from them. Some kinds of parasitic animals, for instance, suck the blood of the host and cause anaemia, from which serious consequences may result. Other kinds steal the host's food, or essential ingredients of it, so that the host suffers from various forms of malnutrition. Others introduce into the host either other kinds of parasitic animals or bacteria, viruses or other injurious agencies. Yet others exert mechanical effects upon vital organs of the host; they may, for instance, exert pressure upon these or block up vital channels, such as blood and lymphatic vessels. By other species vital organs of the host may be severely damaged or destroyed. Its reproductive organs, for instance, may be so injured that the host's biology, and sometimes its sexual characteristics, are altered.

Although the two categories of effects upon the host just mentioned are useful for descriptive purposes, it must be realised that they are quite artificial and that one or more of the processes included in each category may be proceeding at the same time. The sheep stomachworm described in Chapter 3, for instance, usually provokes a chronic inflammatory tissue reaction of the stomach wall and, at the same time, a form of acquired resistance; in addition to these defence reactions, it sucks blood and causes bleeding, so that the host may also suffer from severe anaemia.

We have also to realise that the effects of any particular species upon particular hosts will be influenced by several factors, among which are:

(i) The number and virulence of the parasitic animals and the ability of some Protozoa to multiply their numbers rapidly inside the host, so that their effects can be rapidly increased.

(ii) The susceptibility of the host and its ability or failure to develop resistance (immunity) to the parasitic animal. This question is discussed later in this chapter.

(iii) The kind of tissue affected by the parasitic animal. Some tissues or organs are so vital to the health or life of the host that relatively slight damage done to them will produce a greater general effect than more severe damage done to tissues or organs which are less important.

In general, moreover, it will be true that ectoparasitic animals, especially if they are only temporarily parasitic, will provoke different and less severe reactions than those which affect vital organs such as the heart, liver or kidney. But some ectoparasitic species, especially when they are numerous, can remove considerable quantities of blood from their hosts, so that a severe anaemia is produced. The response of the host's blood-forming organs must therefore be included in the total reaction made by the host. Species which live inside individual cells (*intracellular species*) often do more harm, and do it more rapidly, than those which live between the cells (*extracellular* or *intercellular species*), because they feed upon and destroy the cells which they inhabit. To enter these cells they must be small, and they are usually Protozoa which can, like bacteria, multiply their numbers rapidly, so that their total effects may cause severe and often fatal reactions of the tissues which they attack. Some very small species of parasitic Protozoa are parasitic, not in the cytoplasm of individual cells, but in the nuclei without which these cells cannot live. An example of these is *Cyclospora caryolytica*, which is parasitic in the nuclei of the cells lining the intestine of the mole.

TISSUE REACTIONS AND RESISTANCE PROVOKED BY PARASITIC ANIMALS

It will not be possible, in the short space available in this book, to give more than a very brief outline of the tissue reactions and resistance provoked by parasitic animals. They are essentially the same as the tissue reactions and resistance provoked by bacteria and viruses, and these are fully discussed in text-books of pathology, in which the reader who is interested in this aspect of the host-parasite relationship will find further information about them and full descriptions of the biological processes upon which they depend. He will find in these books, for instance, explanations of the process called inflammation, some form of which frequently occurs when parasitic animals make contacts with the tissues of their hosts; and of the methods by which invading organisms of all kinds are walled off and often destroyed. He will also find in these books full descriptions of the methods by which damaged tissues are repaired, of the consequences of the long-continued irritation of the host's tissues which parasitic animals often cause, of the derangements of the growth of tissues which they may provoke, and of the interaction of antigens and antibodies upon which the various forms of resistance (immunity) depend. In this book it must be assumed that the reader already has this basic knowledge, and we must confine ourselves to very brief references to the tissue reactions which the parasitologist most frequently encounters and to a much-compressed outline of the fascinating and important immunological phenomena which he has to study.

A. TISSUE REACTIONS CAUSED BY PARASITIC ANIMALS

These may be divided into:
  (1) Inflammatory reactions;
  (2) Derangements of normal growth.

(1) *Inflammatory reactions*

These, like the inflammations caused by bacteria, viruses and inanimate agencies, may be acute or chronic. They are the result of some form of injury or irritation caused either by such organs as the teeth of the parasitic animal, or the suckers by which it attaches itself to the tissues of the host, or by its migration through these tissues, or by chemical substances which it secretes or excretes into the host. Frequently it is difficult or impossible to distinguish between direct injury and irritation, because parasitic animals often inflict small injuries which are repeated over periods of time, so that the total effect is one of irritation. The suckers of tapeworms and flukes, for instance, may irritate the host's tissues in this way, so that the characteristic result of continued irritation, namely, chronic inflammation, occurs. The human hookworm, on the other hand, inflicts a relatively greater degree of direct injury upon the wall of the small intestine in which it lives, because it actually engulfs portions of this wall, and detaches and consumes small pieces of it, so that small ulcers are produced. The contents of the food canal may then irritate these ulcerated areas and bacteria may invade them, so that the consequences of bacterial infection may be added to those of the injuries inflicted by the hookworms. In addition to the actual changes in the host's tissues produced by irritation, other effects of it, which are not confined to the area in which it is inflicted, may occur. One example of these must suffice.

Most of the roundworms parasitic in the food canals of their hosts injure and irritate the walls of these canals to a greater or less degree, and the injury or irritation may stimulate the nerves and muscles of the walls of the food canal, so that its normal movements, which propel the food along it, may be increased or disordered. Abnormal reactions, such as vomiting or diarrhoea, may then result. These are unfavourable to the parasitic animal, because they cause the expulsion of numbers of its species from the host. They do not do the host any good either, because they expel, with the parasitic animals, food which helps to maintain the host's health and resistance to the parasitic animal. The expulsion of numbers of the parasitic animals in this manner may, nevertheless, enable the host to get the upper hand; but if the parasitic animal belongs to a species which can multiply its numbers inside the host—if, for instance, it is a protozoon—enough individuals of its species may remain inside the host to enable it to renew its attack, and the host, debilitated by the diarrhoea, vomiting or other reactions of this kind caused by the first attack, may be unable to resist this second multiplication of the parasitic animal.

The visible signs of the effects of inflammation upon the cells of the inflamed tissues are the same whether the inflammation is produced by parasitic animals, bacteria, viruses or inanimate agencies. Those which parasitologists most often encounter are:

(*a*) *Cloudy swelling* (parenchymatous or albuminous degeneration). The cells of tissues affected by this change look pale and swollen and may contain albuminous or fatty granules. Their nuclei may be indistinct. Cloudy swelling is most often seen in organs composed of cells which have a specialised function, such as the liver, kidney or heart muscle. It is usually caused by poisonous substances passed into the host. If its cause is not removed, it may proceed to fatty degeneration and death of the cell.

(*b*) *Fatty degeneration.* When fat appears in cells which do not normally contain it, this is evidence that fatty degeneration is occurring. It is most often seen in the liver and, when the amount of fat thus deposited in the cells is considerable, the tissues affected look pale and yellowish and may feel greasy.

(*c*) *Necrosis.* If any kind of cell degeneration proceeds far enough, the cell dies and this condition of death is called necrosis. If large areas of tissue die, the dead area may look pale and opaque, because the contents of the cells and the tissue fluids in it have clotted. This condition is called *coagulation necrosis*. It occurs especially in cellular organs, such as the liver and kidney.

In some other organs the dead tissue shows softening rather than coagulation, because the process called self-digestion (*autolysis*) tends to occur. Microscopic examination of necrotic areas may reveal some outlines of dead cells, but the area looks granular or structureless, although fragments of cell nuclei may still be visible.

A form of necrosis which occurs in tubercular and syphilitic abscesses is called *caseation* and some species of parasitic animals may also cause it. Caseation often follows fatty degeneration. Tissues which undergo it are slowly transformed into a dry, brittle and granular material, whose consistency resembles that of cheese, although it is more brittle.

Toxic substances secreted by the migratory larvae of *Ascaris lumbricoides*, when they pass through the liver on their way to the lungs, may cause, in the liver cells around them, cloudy swelling, fatty degeneration and necrosis. Numerous macrophages and eosinophil microphages migrate into the tissues surrounding them. The host may successfully limit the damage done by enclosing the parasite in a fibrous capsule. Within this capsule caseation of the damaged tissue and also of the tissues of the parasite eventually occurs, and *calcification* may follow upon this.

This kind of reaction, consisting of a necrotic centre, in which the parasitic animal which causes it, or fragments of it, may or may not be found, is frequently the result of successful reactions of the host against parasitic roundworms, tapeworms and flukes. It is often called a *worm nodule*. It resembles the kind of reaction produced by some forms of chronic inflammation caused by certain kinds of bacteria and may especially resemble the kind of chronic inflammations produced by the bacillus which causes tuberculosis and by the spirochaete which causes syphilis. This resemblance may have considerable practical importance, especially when these types of reaction have gone so far that all traces of the helminths or of the bacteria which caused them have disappeared. The structure of the reaction may then be insufficient by itself to tell us what its cause has been. It is, moreover, possible that some parasitic animals may settle in early tubercular lesions or that the damage which they cause may favour infections of the damaged area by tubercle bacilli. Equally important is the parallel fact that the existence of tuberculosis, syphilis or even of other bacterial infections or of malaria, in a subject infested by roundworms, may render serological tests for the presence of these worms unreliable. This may happen because the cross-reactions mentioned in Chapter 8 may occur.

(2) *Derangements of normal growth*

Among these are the processes which are called *hyperplasia, metaplasia* and *neoplasia*.

(a) *Hyperplasia* is the name given to changes in a tissue which result when the rate of cell division is increased, so that an abnormal increase in the number of cells of that tissue occurs. The size of these cells is not increased. Any enlargement of the tissue or organ undergoing hyperplasia is due to increase in the number of cells in it. Hyperplasia must therefore be distinguished from *hypertrophy*, which is an increase in the size of the cells or of the whole organ which they compose and not an increase in the number of cells present.

Among parasitic animals which cause hyperplasia of the organs in which they live are the protozoon *Eimeria stiedae*, when it infects the liver of the rabbit, the fluke *Fasciola hepatica*, parasitic in the liver, and the human blood fluke, *Schistosoma haematobium*, the spined eggs of which may cause hyperplasia of the human bladder. Hyperplasia may be accompanied by metaplasia next to be described.

(*b*) *Metaplasia* is a term which means a transformation. It occurs when there is a change in the type of cell present in the part of the body affected. It may accompany hyperplasia and is a less common result of the effects of parasitic animals than the other growth changes just mentioned.

An example of it is the change of the normal cylindrical cells which line the smaller air tubes (*bronchioles*) of human beings and carnivores into layers of cells similar in shape to those of mammalian skin, when the lung tissue in the neighbourhood of these bronchioles is infected with the fluke *Paragonimus westermanii*.

(*c*) *Neoplasia* means the formation of tumours. A tumour is a new growth from existing tissue, which serves no useful purpose, usually performs no function and follows its own laws of growth independently of the laws which govern the growth of the organism in which the tumour occurs; this organism may, in fact, be severely damaged by the tumour, or may be killed by it. Tumours are sometimes classified into the two groups, simple or benign tumours and malignant tumours.

Simple or benign tumours do not reproduce themselves elsewhere in the organism in which they occur; their effects are less serious than those of malignant tumours, although they may exert, by their pressure and other influences upon the tissues which surround them, serious effects.

Malignant tumours may be reproduced elsewhere in the organism in which they occur. Portions of them may be detached and carried by the blood stream to other parts of the organism, in which they settle down and grow, so that secondary tumours are produced composed of the same kind of tissue, which are called *metastases*. The effects of malignant tumours are always serious and usually they are fatal.

The causes of tumour formation cannot be discussed here. We must be content with the following instances of tumours caused by parasitic animals.

The hyperplasia caused by the blood fluke, *Schistosoma mansoni*, may result in the formation of the kind of tumour which is called a *papilloma*. A papilloma is a tumour which consists of a central core of connective tissue containing blood vessels covered with epithelial cells, and often it takes the form of finger-like outgrowths, which may branch to form cauliflower-like masses.

Papillomas may be contrasted with the kind of tumour which is called an *adenoma*. An adenoma consists of the secretory cells of a gland bound together by fibrous tissue. It arises mainly in secretory glands and may contain cavities full of secretions, called cysts. The growths caused in the liver by *Eimeria stiedae* are adenomata, although they are sometimes classified as papillomata. Other instances of tumour-like growths caused by parasitic animals are the growth in the walls of the stomach of horses caused by the roundworm *Habronema megastoma* (see also Chapters 6 and 9), and those caused in the stomach wall of the cat, leopard and dog in India by the roundworm *Gnathostoma spinigerum* (see also Chapter 9), whose life history resembles that of *Diphyllobothrium latum*.

Some parasitic helminths may cause the formation of tumours which tend to become cancerous. Some species of roundworms (lungworms) parasitic in the lungs of sheep, for example, especially those belonging to the family *Protostrongylidae*, may cause the formation of tumours which are benign when they begin, but later become cancerous.

Cancerous growths in the stomach and tongue of the rat have been produced experimentally by feeding them with cockroaches infected with the larvae of the roundworm *Gongylonema neoplasticum*; and it is stated that *Cysticercus fasciolaris*, the bladderworm stage of the tapeworm, *Taenia taeniaeformis*, the adult stage of which is parasitic in the small intestine of the cat, stoat and other carnivores, may cause cancer of the livers of rats, mice and other rodents in which *Cysticercus fasciolaris* occurs. Some authorities have associated hydatid cysts of the lungs, which are the larval bladderworm stages of the tapeworm, *Echinococcus granulosus*, with the formation of cancer of the lungs. Cancerous growths may be caused in the liver of man by the flukes *Opisthorchis (Clonorchis) sinensis* and *O. felineus*; and in the human bladder by the blood fluke, *Schistosoma haematobium*. The fluke *Paragonimus westermanii* may cause cancer of the lung of the tiger.

B. Resistance (Immunity) Provoked by Parasitic Animals

Because the resistance provoked by parasitic animals is essentially the same as that provoked by bacteria and other agents, parasitologists and bacteriologists classify the various kinds of resistance under the same headings. In this book they will be grouped into two categories, namely,

Innate Resistance, which is often called *Natural Immunity*, and *Acquired Resistance*, which is often called *Acquired Immunity*.

The term resistance is used instead of the term immunity, because it expresses better the fact that this kind of reaction does not always produce that complete freedom from the effects of the invading organisms which the word immunity implies.

It will be useful also to note here the distinction which parasitologists draw between *normal* or *usual hosts* and *abnormal* or *unusual hosts*. Normal hosts are the ones in which particular species of parasitic animals are usually found. Sheep and cattle are, for instance, normal hosts of the liver fluke, *Fasciola hepatica*. But many species of parasitic animals are also found, under certain circumstances, in hosts in whose bodies they can become parasitic, although they do not usually do so. These latter hosts are the abnormal or unusual hosts of these species. Man is, for instance, one of the unusual hosts of the liver fluke, *F. hepatica*.

(1) *Innate Resistance to Parasitic Animals* (*Natural Immunity*)

This is sometimes called natural or genetic resistance.

A host shows innate resistance to a particular species of parasitic animal when that species has the opportunity to infect the innately resistant host, but cannot become parasitic in it. Man may, for instance, swallow the oocysts of *Eimeria caviae*, but this species cannot become parasitic in him. Pasteur found that some individual silkworms are innately resistant to the protozoon *Nosema bombycis*, which causes silkworm disease, and he bred from these individuals a naturally immune strain of silkworms. It has been found that some individual gnats belonging to the species *Culex pipiens* are innately immune to the malarial parasite of sparrows and other birds, *Plasmodium cathemerium*, of which they are the definitive hosts, and that their innate resistance behaves as a Mendelian dominant when the gnats are bred. When a parasitic animal requires an intermediate host, the near relatives of that intermediate host may be innately resistant to the parasitic animal. Thus some species of snails are innately resistant to the miracidial larvae of flukes which infect their relatives. The practical importance of this fact is indicated below.

When a particular species of parasitic animal thus fails to become parasitic in a particular species of host, that host's innate resistance is said to be complete. Frequently innate resistance lasts throughout the lifetime of the host which possesses it. It may, on the other hand, be incomplete. One variety of incomplete innate resistance, namely, age resistance, is discussed at the end of this chapter.

In the category of innate resistance we have also to consider what are called *physiological strains* of certain species of parasitic animals. The large roundworm, *Ascaris lumbricoides*, provides an example of these. If we compare specimens of this species taken from the intestine of man and the pig respectively, we find that we cannot distinguish the one from the other by any feature of their structure. Infective larvae derived from specimens parasitic in the pig will not, however, develop to maturity in man, and those derived from man will not develop to maturity in the pig. Man thus has innate resistance to the pig strain of *A. lumbricoides* and pigs are innately resistant to the human strain of it. The innate resistance of both is, however, incomplete, because the larvae of the pig strain can live long enough in man to enter his lungs and to cause pneumonia and other symptoms of the damage these larvae do to the lungs. The larvae of the human strain can likewise cause signs of damage done by them to the lungs of pigs, provided that the general health, and therefore the general resistance, of the pig is first lowered by administration to it of a diet deficient in vitamin A.

The similar physiological strains of the dog hookworm, *Ancylostoma caninum*, have been mentioned in Chapter 6. One of them lives better in the cat than in the dog. These physiological strains which are structurally alike remind us of some genera of Protozoa, such as *Theileria*, species of which cause East Coast fever and similar diseases of cattle. These species are structurally alike and can be distinguished only by the diseases that they cause,.and by the immunological reactions of their various hosts to them. They resemble the 'types' recognised among bacteria, and we may yet learn that they are not species at all, but biological races within true species, similar to those that are known among the malarial parasites (see Chapter 10).

Innate resistance—and indeed any other kind of resistance—may be broken down by malnutrition, by the influence upon the host's general health of other diseases and by other factors discussed below, and this fact has important bearings on the control of diseases caused by parasitic animals and other agents. Many examples of the breakdown of innate resistance could be given, but we can mention here only the occasional successful invasion of man by the liver fluke, *Fasciola hepatica*, the normal hosts of which are sheep, cattle and some other ruminant animals. Man may also be occasionally infected by the hookworms, *Ancylostoma braziliense* and *A. caninum*, the normal hosts of which are cats and dogs (cf. Chapter 9).

Such occasional breakdowns of innate resistance are important not only because, when they occur, serious diseases may result, but also because they may indicate one way in which parasitic animals may extend the number of species of hosts inside which they may become parasitic. If the innate resistance of a particular host breaks down frequently in an area in which the parasitic animal which can take advantage of the breakdown is abundant and vigorous, the innately immune host which experiences the breakdown may, in the course of evolution, lose its innate immunity and become a normal host of the parasitic animal in question.

This may happen to the intermediate as well as to the definitive host. The sheep liver fluke, *Fasciola hepatica*, for example, has been carried by man all over the world in domesticated animals infected with this species which man has taken with him. In several countries into which it has been introduced, *F. hepatica* has acquired the ability to use several species of snails as its intermediate hosts (see Chapter 4). No doubt some of them were not originally innately resistant to the miracidial larvae of *F. hepatica*, but some of them may have been.

The blood flukes belonging to the genus *Schistosoma* have not, at present, found in the United States of America species of snail which they can use as intermediate hosts. All of the species of snails available in that country appear to be innately immune to infection with the miracidial larvae of these dangerous parasites of man. But the example of *Fasciola hepatica*, which is related to the blood flukes, suggests that the human blood flukes also may yet encounter a breakdown of the innate immunity of United States aquatic snails. If this happened, human blood flukes would be added to the parasitic animals which are endemic in the United States. It is, however, unlikely that there are in the United States enough persons infected with blood flukes to provide sufficient miracidial larvae to establish the infection of unusual snail hosts. It is perhaps more likely that filarial roundworms, aquired by United States soldiers during the Second World War, may find in the United States mosquitoes which are not innately resistant to them and can act as their intermediate hosts.

(2) *Acquired Resistance to Parasitic Animals* *(Acquired Immunity)*

This may be either (*a*) *active*, or (*b*) *passive*. It is active when a host responds by its own efforts to the presence of a parasitic animal in it by producing antibodies which operate against that parasitic animal. It is passive when the host makes no effort of its own to produce the resistance, but passively receives it, either from its mother by way of the placental blood or the milk, or from man, who artificially injects antibodies into it.

Active resistance is more powerful than passive resistance, but usually both are more powerful against parasitic animals which enter the interior of the host's body and affect its tissues than against those which live only on the host's surface or in its food canal.

(*a*) *Active Acquired Resistance*

This may be acquired (i) *naturally* as the result of an infection with the parasitic animal which has occurred in the course of the natural life of the host; or (ii) *artificially*, as a result of the introduction by man of either the living or dead parasitic animal, or of extracts of its substance.

(i) *Naturally acquired active resistance* is shown by some hosts of all of the main categories of parasitic animals.

Cattle naturally infected with *Trypanosoma brucei* or *T. congolense* develop resistance to these species and to the species of *Theileria* mentioned above and to those of *Babesia* (cf. premunition below and Chapters 8 and 9). The resistance thus naturally acquired may be a valuable quality of the cattle. Man develops naturally acquired active resistance to the malarial parasites, *Plasmodium vivax* and *P. knowlesi*, and to *Leishmania tropica* (fig. 19), the cause of Oriental sore; birds develop it to several species of avian malarial parasites; rats develop it to *Trypanosoma lewisi* and various rodents develop it to *Trypanosoma cruzi*. Infections with some species of Coccidia also provoke a resistance to subsequent infections with them. Thus hen chickens infected with *Eimeria tenella* and *E. necatrix* become resistant to these species, rabbits become resistant to *E. perforans*, and cats to *Isospora rivolta* and *I. felis*.

Infections of mammals with flukes rarely provoke resistance, but some kinds of marine fish do develop resistance to flukes which infect them. Among the tapeworms, intermediate hosts which harbour the bladderworms provide the best examples of resistance provoked by natural infection. Thus rabbits become resistant to subsequent infections with the bladderworms of *Taenia pisiformis*, rats to those of *T. taeniaeformis* and oxen to those of *T. saginata*. Complete resistance is rarely, if ever, acquired by the definitive host which harbours the adult tapeworm (cf. premunition below).

Among the hosts of roundworms which develop resistance as a result of natural infections are cats, which resist second infections with the roundworm *Toxocara cati*, and chickens, which resist second infections with the roundworm *Ascaridia lineata*. Partial resistance to subsequent infections with hookworms is, however, acquired by man and dogs, sheep resist later infections with the sheep stomachworm and its relatives (cf. below), and rats resist later infections with the pork trichina-worm.

(ii) *Artificially acquired active resistance* is the result of man's artificial introduction into a host of a living or dead parasitic animal or of extracts (antigens) made from its substance.

Bacteriologists can use this method of producing resistance more effectively and over a much wider range of species than parasitologists can, because they are able to cultivate artificially so many species of bacteria, whereas the parasitologist is able to cultivate artificially only a very few species of parasitic animals, all of which are Protozoa. He must therefore make his antigens from species obtained from natural or experimental infections of the host. Attempts to produce active acquired resistance by this method are, for this reason, limited to relatively few species of parasitic animals. These attempts are, moreover, hampered by the fact that the introduction into the host to be immunised of material derived from one strain of a particular species of a parasitic animal may fail to provoke resistance to other strains within that species. There are, for instance, various strains of trypanosomes which cause disease of man and domesticated animals, and the difficulty of inoculating into man and domesticated animals all these strains so that resistance to all of them may be acquired has defeated all attempts to provide an effective artificial acquired resistance to these trypanosomes.

With some species, however, a degree of success has been attained. The introduction into young cattle of the protozoon, *Babesia bigemina*, has been used successfully to immunise these animals against Texas fever. Young cattle exhibit to this disease the reversed age-resistance described below. The young cattle are, that is to say, more resistant to it than older cattle are, so that when *B. bigemina* is introduced into them they suffer only a mild form of Texas fever, but acquire a resistance to it which protects them from later infections acquired when they are older. Cattle thus immunised are called 'salted' cattle, and this artificially acquired active resistance has saved the lives of many cattle. It is exactly similar to the naturally acquired resistance to this species mentioned above. A similar procedure can be used to immunise the host against other species of *Babesia* and against some of its relatives.

Attempts to produce artificial active acquired resistance to other species of parasitic animals have been less successful. Man, for instance, cannot be protected against his malarial parasites in this way, although birds can be given a relatively low degree of resistance to some species of bird malarial parasites by the introduction of these parasites into them. There is also evidence of partial success in the production of artificially acquired active resistance to some species of Coccidia and some kinds of bladderworms of tapeworms. The hosts of various species of roundworms, such as ascarids, hookworms and the pork trichina-worm, can be given an active acquired resistance by the introduction of these species into them, but the resistance thus provoked has little or no practical value. It has also been claimed that man can be given a resistance to the protozoon *Leishmania tropica* (fig. 19) by the introduction into his skin of living Protozoa of this species. Very small doses of living cultures of *Trypanosoma brucei*, which causes the disease of cattle called nagana, have been used to give cattle a resistance to this disease, but the resistance is not powerful.

The production of artificially acquired active resistance by injection of either dead parasitic animals, parasitic animals which are not virulent, or extracts of dead bodies of parasitic animals is better known. Resistance thus provoked is, however, usually less powerful than resistance produced by infection with the living parasitic animal, either because the species injected into the host is not living and growing and therefore stimulates the host less powerfully, or because the preparation of the extracts of the dead parasitic animals may fail to retain the antigens whose action on the host is necessary for the production of the antibodies upon which the resistance depends.

Complete resistance of rats to *T. lewisi* can, nevertheless, be provoked by vaccination of them with killed cultures or with suspensions of this species of trypanosome. Killed cultures of trypanosomes which cause disease, on the other hand, produce a much feebler resistance or none at all, so that they cannot be used for the treatment of infections with these species. The inoculation of killed Coccidia fails to provoke resistance of the host, and killed malarial parasites of man and monkeys also fail to produce it. Canaries, however, which are vaccinated with bird malaria parasites which have been either killed or made non-virulent by exposure to a low temperature, do become comparatively resistant.

Experiments with the dried body substance of flukes, tape-worms and roundworms have produced interesting results. Thus rabbits are made partially resistant to the sheep liver fluke, *Fasciola hepatica*, by injection into them of the dried substance of this fluke. The injection of the dried substance of hydatid cysts will make sheep resistant to infection with these larval forms of *Echinococcus granulosus*. Injection of the dried substance of the pork trichina-worm, *Trichinella spiralis*, will make rats resistant to this species, and they may be made resistant to a few other species of roundworms in this way. The use of the dead bodies of parasitic animals, or of extracts of them is not, however, at present useful for the treatment of parasitic diseases.

Whether it is naturally or artificially acquired, the result of active acquired resistance may be that, when the resistant host is infected by the parasitic animal, it holds the parasite in check so that, although the parasitic animal may be doing the host considerable harm, it does not seriously affect the host's health. The host and the parasite have established an equilibrium which enables both to live. The host is exhibiting what is called *tolerance* of the parasitic animal and it is a source of the infection of other hosts. It is, as we say, a *carrier* of the parasitic animal. When the parasitic animal has become so well adapted to its host that it practically never causes disease in that host, but may, when it gets into other hosts which are less tolerant of it, cause disease, the tolerant host is called a *reservoir host*, because it acts as a reservoir from which other animals may be infected. The discovery and control of reservoir hosts of species of parasitic animals which cause serious diseases of man and his domesticated animals is obviously one of the most important tasks of the parasitologist who wishes to control disease.

In the tropics, for instance, large numbers of people are carriers of the amoeba which cause amoebic dysentery of

man, and in Europe this amoeba is by no means uncom-
mon. Examinations of British people have shown, for
instance, that 3–10 % of the population examined are
infected with it. About half of these people are infected
with a small race of the amoeba which never causes
disease. The other half are carriers of the large race of it,
which causes amoebic dysentery in only about 10 % of
people infected with it. Why it does so we do not yet know,
but it is evident that, even in England, there are some
thousands of people who are, without any suspicion of the
fact, capable of infecting other people with either the
harmless or the potentially pathogenic race of this amoeba.
Most, if not all, human infections with it come from other
human beings, but natural infections are also found in
some kinds of monkeys and in kittens, dogs and pigs. Man
himself is therefore the most important reservoir host of
this species.

He is also the most important reservoir host of the
human hookworms. *Ancylostoma duodenale* has been found
only in man, some species of monkeys, wild carnivores and
occasionally in the pig. *Necator americanus* has been found
only in man, some kinds of monkeys and at least once in
the rhinoceros, pangolin, and one species of rodent (*Coendu
villosus*). This information tells us that, apart from man
himself, only the animals mentioned are possible sources of
hookworm disease of man. In practice the information
amounts to the fact that man himself is the usual source of
his own hookworm infections (cf. Chapter 10).

On the other hand, we know that although two of the
species of blood flukes which cause serious disease of man,
namely, *Schistosoma haematobium* and *S. mansoni*, are found
only in man and some species of monkeys, the other species

which causes human disease in the Japanese and Chinese areas of the Far East, *S. japonicum*, is found, not only in man, but also in dogs, cats, rats, mice, field-mice, cattle, water buffalo and horses, all of which are reservoir hosts of this species of parasitic animal from which it can be transmitted to man by means of its cercarial larva, which can penetrate the human skin.

Another interesting example of the reservoir host has been revealed by Russian work on leishmaniasis of human skin. This disease, which takes the form of ulcers of the skin called tropical sores or Bagdad boils, is caused by infections with the protozoon, *Leishmania tropica*, which is transmitted to man by sandflies belonging to the genus *Phlebotomus*. In Turkestan two forms of this disease are recognised, a dry and a moist type. The moist type affects man only during the summer and autumn.

The reason why this is so is that certain wild rodents living in the desert regions of Middle Asia, especially some species of gerbils and sousliks, are naturally infected with *Leishmania tropica* (fig. 19), and sandflies belonging to the genus *Phlebotomus*, which transmit *Leishmania tropica* to man, also transmit it among the rodents. The sandflies share the deep burrows of the rodents, in which they find shelter during the severe winters of these regions; they live and breed in these burrows the whole year round and maintain the infection among the rodents throughout the year. When, during the warmer months, the weather is suitable for them, the sandflies emerge from the rodents' burrows and transmit *L. tropica* to man.

Attacks upon the sandflies have considerably reduced the numbers of human beings infected with this moist type of sore. It seems likely that the rodents are the usual hosts of this species of parasitic animal, while man is probably an unusual host of it, although he may by now have become practically a normal host.

It used to be thought that *Trypanosoma gambiense*, the flagellate protozoon which causes African sleeping sickness of man, could be passed to man from African antelope and other ruminants in that country; but there is no proof that this actually happens. It can live in the blood of antelope, sheep, goats and pigs and may possibly be transmitted by tsetse flies from these animals to man; but usually human infections are acquired from other human beings. *T. rhodesiense*, on the other hand, the cause of sleeping sickness in Rhodesia and some other parts of East Africa, does exist naturally in wild game and domesticated animals, which are therefore reservoir hosts from which it is transmitted to man by tsetse flies. *T. gambiense* has apparently become adapted to life in the blood of man, while *T. rhodesiense* is only now achieving that adaptation. Both species appear to have been derived from *T. brucei*.

The form of resistance, to which the name *premunition* is given, requires explanation here. The term means active, acquired resistance, acquired either naturally or artificially, which persists only so long as the host remains infected with the parasitic animal which provokes the resistance. Premunition therefore differs from other forms of resistance, which remain after the parasitic animals that provoke them have left the host or have died.

African cattle, for instance, so long as they are infected with *T. brucei*, are resistant to this trypanosome and the number of trypanosomes able to live in the resistant cattle is restricted. When, however, the trypanosomes die, or are killed by drugs given by man, the resistance of the cattle disappears and they become susceptible to sub-sequent infections with *T. brucei*. The resistance of cattle to *Babesia* mentioned below is another instance of pre-munition, because the babesias persist in the blood while the cattle are resistant to them. The lifelong resistance of cattle to *Theileria mutans*, *T. annulata* and *T. dispar* is also premunition, because these species of Protozoa also persist in the blood of the cattle while the resistance lasts. The resistance of cattle to *T. parva*, however, is not premunition, because *T. parva* never persists in the blood of cattle which survive the disease (East Coast fever) caused by this species and thus become resistant to it.

Another instance of premunition is the resistance of birds and monkeys to avian malarial parasites which limits the number of these parasites able to live in the birds and monkeys, but disappears when the infections cease.

Some experts believe that the greater resistance of the negro race to human malarial parasites (see below) is really premunition, being the result of a continuous infection with these parasites which begins early in life and is maintained by repeated subsequent infections.

The instances of premunition just given are all produced by parasitic animals which can, once they have entered a host, multiply in it. When the parasitic animal is a fluke, tapeworm or roundworm, it cannot multiply like this. The number of individual parasitic animals in the host can then be increased only by the entry of more infective phases. When premunition against these kinds of parasitic animals develops, only a certain number of the infective phases which enter it can survive and grow up to maturity. The growth of all the others is retarded, they cannot attain sexual maturity, and many of them are expelled or die inside the host.

This phenomenon is called *resistance to further infection*. Like the other forms of premunition just described, it can be produced by either natural or artificial infection. It is well shown by certain hosts of roundworms, such as sheep infected with the sheep stomachworm and related species of roundworms.

The precise nature of the resistance caused by infection with some species of roundworms, tapeworms and flukes is, however, still uncertain. Some of the examples given above of naturally acquired resistance may eventually prove to be instances of premunition which persist only so long as the infection exists. Certainly man seems to show premunition to the beef tapeworm, *Taenia saginata*, and to the pork tapeworm, *T. solium*, because usually only one individual tapeworm of either of these species can survive in him. Either can, however, live in the human intestine together with the fish tapeworm, *Diphyllobothrium latum*. The fish tapeworm is not necessarily solitary. As many as 90 individuals of it have been found in a single human being. The occurrence of several individuals of both *T. solium* and *T. saginata* in the human intestine at the same time has, however, been recorded, and children, especially if they are badly nourished or otherwise debilitated, may harbour thousands of individual tapeworms of the species *Hymenolepis nana*. Poultry likewise may harbour many individual tapeworms of the genus *Davainea*.

(b) Passive Acquired Resistance

This is never the result of the host's own activity. It is always the result of the introduction into a particular individual of antibodies made by another host of the same or of a different species. It is usually weaker than active acquired resistance and it does not last so long. Usually its duration is reckoned in weeks or a few months, rather than in the years during which some kinds of actively acquired resistance may last. Like active acquired resistance, it may be acquired (i) *naturally* or (ii) *artificially*.

(i) Passive resistance is *naturally acquired* only by young animals from their mothers who are resistant. The mother suffers from some kind of infection, makes antibodies against it and transmits these antibodies to the young. She can do this by three routes, by the placental blood stream, by the colostrum which nourishes the offspring before the mother's milk is available, or by the mother's milk itself.

(*a*) When passive resistance is transmitted by the mother to the young by way of the placental blood stream, the young are made resistant before they are born. We have, at present, little evidence of the transference of passive resistance to parasitic animals in this manner, but it has been shown that mice infected with the tapeworm *Hymenolepis nana* var. *fraterna* can transmit resistance to this species to their young while these are still in the uterus, but that the resistance thus transmitted lasts only 9 days and is less powerful than resistance to the same tapeworm that is transmitted by the mother's milk (see below).

(*b*) Transmission of passive resistance by the mother to the young by way of the colostrum is an interesting biological phenomenon which gives valuable protection to many young animals. During the first few hours of their lives after birth, they are suckled, not on the mother's milk, but on a secretion of the mother's mammary glands, called the colostrum, which contains a rich store of antibodies present in the mother's body. Because the digestive juices of the newly born young are not yet active, this colostrum and the nutrient material and antibodies in it are rapidly taken into the body of the young animal, so that it becomes resistant very soon after its birth. The resistance thus conferred on it, like all forms of passive resistance, does not last long, but it protects the young at a time when they are very vulnerable. Most of the known instances of resistance conferred on the young in this manner are instances of resistance to bacteria, but there is some evidence that resistance to a few species of parasitic animals may also be transferred in this way. It has been shown that artificial injection into the blood of calves of the colostrum of cattle already resistant to the protozoon *Trichomonas foetus* confers on the calves a certain degree of resistance to this parasite, so that presumably resistance can also be naturally conferred by the mother cow. Mother rats which are resistant to *Trypanosoma lewisi* and *T. cruzi*, can transmit their resistance to these species to their young within 24 hours of their birth. By it the young are completely protected. It is doubtful, on the other hand, whether resistance to these species can be transmitted through the placental blood.

(c) Transmission of passive resistance by way of the mother's milk does not, naturally, protect the young so soon after their birth as the resistance transmitted by the colostrum does. Antibodies transmitted in the mother's milk must, moreover, encounter the digestive ferments of the young, which are active by the time that the mother's milk is being taken in; and these ferments may alter or destroy the antibodies. A degree of resistance may, nevertheless, be transmitted in this way. Like the forms of passive resistance transmitted through the placental blood and colostrum, it is weaker than active resistance and lasts usually only a few weeks. Young animals may be protected by their mother's milk in this way against both bacteria and parasitic animals, but instances of immunisation by the mother's milk against parasitic animals are relatively few.

It is known, however, that young rats may acquire antibodies against *T. lewisi* and *T. cruzi* from their mother's milk. This may happen even when the mother is artificially infected with these trypanosomes after her young have been born, so that, so long as the mother is producing milk, she can protect her young against infection with these trypanosomes. Mice resistant to *T. duttoni* are also able to transmit their resistance to this species to their young by way of the milk. Instances of the transference of resistance to other kinds of parasitic animals in this manner are rare, but female mice can transmit in this way a degree of resistance to the tapeworm *Hymenolepis nana* var. *fraterna* and female rats, rabbits and hamsters which are resistant to the pork trichina-worm, *Trichinella spiralis*, can transmit a resistance to their young which is, it is believed, transmitted through the milk. It lasts for about the first three weeks of the life of the young.

(ii) Passive resistance is *artificially acquired* when man injects into an animal antibodies made by some other animal. When, for instance, the *Pneumococcus* is injected into a rabbit the rabbit develops in its blood antibodies which make it resistant to this coccus. If its serum is injected into a rabbit not resistant to the *Pneumococcus*, resistance to this coccus is passively and artificially transferred to the previously non-resistant rabbit. Artificially acquired passive resistance thus takes effect immediately the resistant serum is injected. It is always specific, because man injects only the antibodies which operate against the organisms against which he wishes to create a passive resistance.

This fact, that artificially acquired passive resistance must be specific, limits its use against parasitic animals, because all forms of resistance to parasitic animals are less specific than the forms of resistance to bacteria. Artificial transference of passive resistance to a few species of parasitic animals has, however, succeeded.

Rats, for instance, can be made resistant to *Trypanosoma lewisi* and *T. duttoni* by injection into them of the serum of other rats already resistant to these species; and rats infected with *T. cruzi* can be treated with the sera of animals made resistant to this species. Human malaria cannot be treated with the serum of human beings suffering from malaria, but some species of the malarial parasites of monkeys can produce resistant sera which will, when they are injected into non-resistant monkeys, reduce their infection with these species. Neither infections with Coccidia, *Entamoeba histolytica* of man nor flukes can, so far as we know, be influenced by sera of hosts resistant to these species. The sera of rats and rabbits infected with the bladderworms of some species of tapeworms will, however, protect non-resistant rats against these bladderworms. The sera of rats already resistant to *Trichinella spiralis* and those of hen chickens already resistant to *Ascaridia galli* protect non-resistant rats and chickens respectively against these two species.

THE NATURE OF INNATE AND ACQUIRED RESISTANCE TO THE EFFECTS OF PARASITIC ANIMALS

The nature of passive resistance is obvious. It is due to the antibodies transferred by the mother to her young or injected by man. The nature of innate and active acquired resistance to the effects of parasitic animals is much less clearly understood.

The Nature of Innate Resistance

It has been suggested that *innate resistance* to some species of parasitic animal is an inherent incompatibility of the host and the parasitic animal which has been developed in the course of evolution and that it is inherited. This may be so; but we wish to know the precise factors upon which this inherent incompatibility depends. Existing knowledge suggests that certain instances of it are due to chemical and physical factors in the environments of the parasitic animals. The infective larvae of species of roundworms which are normally parasitic in the food canals of sheep cannot, for example, survive in the food canals of carnivorous animals, such as the dog, nor in the alimentary canal of man. The reason for this may simply be that these larval infective phases cannot survive the action of the powerful digestive juices of man and of carnivorous animals like the dog and are therefore unable, at present at any rate, to become parasitic in them.

An example of the poisonous action of a normal physiological secretion of the host which excludes in this way infective larval stages of some species of parasitic animals is provided by the experimental work which has been done upon the bladderworm, *Cysticercus pisiformis*, which is the bladderworm stage of the tapeworm, *Taenia pisiformis* (*T. serrata*). It has been claimed that sodium glycocholate kills the heads of the tapeworms in the bladderworms of this species, so that hosts whose bile normally contains sodium glycocholate cannot become infected by *T. pisiformis*. The bile of the dog, however, does not contain sodium glycocholate, so that *T. pisiformis* can establish itself in the dog.

The Nature of Active Acquired Resistance

There is now no longer any doubt that this is due to the production by the host of specific antibodies, and that its nature is not essentially different from that of acquired immunity to bacterial infections. Because it is possible to prepare antigens from parasitic animals and to perform with them and with the sera of hosts of parasitic animals the principal serological tests normally used by immunologists, the serum of these infected hosts must contain antibodies which operate against the parasitic animals in question.

Factors which Influence Resistance to Parasitic Animals

These may be divided into factors which influence the host and factors which the parasitic animal brings into operation.

The host's resistance will be influenced by its *susceptibility* to the parasitic animal and by the *number* and *virulence of the parasitic animals* which invade it. Small numbers of a virulent species may more readily overcome the host's resistance than relatively large numbers of a less virulent species. The virulence of parasitic animals is, we have already learned, generally greater in unusual than in usual hosts of the parasitic animal, the latter having adapted themselves to the effects of the parasitic animal. The influence of the number and virulence of parasitic animals is, perhaps, more characteristic of the single-celled Protozoa, which multiply their numbers in their hosts much as bacteria do, than of the helminths, the adult stages of which do not multiply. In their intermediate hosts, however, some species of helminths multiply and this multiplication must have an important influence on the resistance of the intermediate host to them.

The susceptibility of the host may be increased, and its resistance may be modified or broken down, by any influence which undermines the health of the host to a sufficient degree (see also Chapter 8). Important among these debilitating influences is the existence of other diseases. These may be caused by bacteria or other agents; or they may be caused by species of parasitic animals different from the species which is provoking the resistance under consideration. A man, for example, who has been rendered anaemic by hookworm disease will suffer more severely from malaria than a healthy man will. The operation in the reverse direction of this reciprocal effect of diseases caused by bacteria and parasitic animals has already been mentioned. Many sheep, for example, are debilitated by the effects of roundworms parasitic in their intestines and for this reason more readily acquire bacterial infections which kill them.

Even more important than the influence of other diseases is *malnutrition*. Malnutrition is not always easily detected. It is, in some parts of the world, widespread among human beings and their domesticated animals. Some medical men have, indeed, described it as the most important of the tropical diseases of man. It must be one of the main causes of the breakdown of human resistance to parasitic animals, and certainly it is one of the main causes of the breakdown of the resistance of our domesticated animals to them.

Much work has been done with the object of finding out which particular elements of the diet have the greatest influence upon resistance to the effects of particular species of parasitic animals. This work cannot be summarized here, but certainly some of the vitamins are necessary if adequate resistance is to be maintained.

Experimental work has, for example, shown that the resistance of poultry to infestations with the roundworm, *Ascaridia galli*, is decreased when the diet lacks vitamins A and D and the vitamins of the B complex, and also when it does not contain enough protein. An inadequate supply of certain minerals may also reduce the resistance of animals to infestation with some species of parasitic animals. Iron is especially required by the hosts of parasitic animals which suck blood and thus cause anaemia. Without an adequate supply of it in the diet, the organs which make the iron-containing haemoglobin removed from the blood, or the haemoglobin altered by parasitic animals, cannot replace this essential substance. There is some evidence also that relatively minute quantities of cobalt, and perhaps also of copper and manganese, help sheep and cattle to establish and maintain resistance to infestations with certain roundworms.

Another factor which may influence the resistance of individual hosts to parasitic animals is their *age*. Earlier in this chapter it was explained that parasitic animals establish themselves more easily in young hosts than in older ones. This age factor is important. Many losses of farm animals, for example, are losses of young animals. Later on, the young host, if it survives, develops a resistance which protects it in greater or less degree. The farmer seeks, therefore, to nurse his young animals through this dangerous young phase of their life, during which they are more susceptible to the entry of parasitic animals. This later development of resistance to the parasitic animal by the host when it becomes older is sometimes called *age-resistance (age-immunity)*.

True age-resistance exists only when young animals, which have not been previously infected with a particular species of parasitic animal, are more easily infected when they are young than when they are older. It may be shown by either the usual or the unusual hosts of the parasitic animals concerned. It must, of course, be clearly differentiated from the transmission of the resistance of the mother to the young described above under the heading of passive resistance. When a young animal has been passively made resistant by its mother, it is not truly age-resistant.

Parasitic protozoa against which age-resistance may develop include some species of them which attack man. Thus human children are more susceptible to the malarial parasite, *Plasmodium vivax*, which causes benign tertian malaria. *Entamoeba histolytica*, which causes human amoebic dysentery, most often attacks young human adults and infections are not often acquired by people older than 35, although in certain places, in which circumstances favour infection, many aged people may acquire infection with this amoeba. Young dogs and cats may also suffer severely from infection with it.

Age-resistance of man and other animals to *Trypanosoma*

cruzi has often been reported, and young rats are more susceptible to *T. lewisi*. Young rats are, on the other hand, more, not less, resistant to *T. equiperdum*. In other words, their age-resistance to this species is reversed. The fact that cattle show a similar reversal of age-resistance to the protozoon *Babesia bigemina*, to which young cattle are more resistant than older ones, has been mentioned earlier in this chapter. Another instance of the reversal of age-resistance is the greater resistance of young sheep to the lungworm, *Muellerius capillaris*, which is rarely found in lambs less than 6 months old. In the same way young hen chickens are more resistant to the coccidian *Eimeria necatrix* than older ones are, but, under the age of 3 months, they are less resistant to *E. tenella*.

There is little evidence of age-resistance to flukes, although human schistosomiasis is acquired more often between the ages of 10 and 15 years, and few people over 30 years old are infected. Similarly, the tapeworm *Hymenolepis nana* affects children aged 5–14 years most often, and rats are more resistant to the related species *H. fraterna* when they have reached the age of 5–7 months.

There are many instances of age-resistance to parasitic roundworms. The pork trichina-worm is more readily established in young dogs than in older ones, in which its larvae fail to mature. Age-resistance is also shown to: the strains of *Ascaris lumbricoides* which infect man and pigs respectively; *Ascaridia galli*, parasitic in the small intestine of poultry; *Ancylostoma caninum*, parasitic in the small intestine of the dog, and *A. braziliense*, parasitic in the small intestine of the cat; the pinworm, *Enterobius vermicularis*, parasitic in the intestine of man; *Oesophagostomum columbianum*, parasitic in the intestine of sheep; the lungworms, *Dictyocaulus filaria* and *Protostrongylus rufescens*, parasitic in the bronchi of sheep; *Haemonchus contortus*, parasitic in the true stomach (abomasum) of sheep. These are all the usual hosts of these roundworms. Examples of age-resistance of unusual hosts are: the dog, when it is infected with the human hookworm, *Ancylostoma duodenale*, and the young hen chicken when it is infected with the worm, *Syngamus trachea*, the normal host of which is, in the opinion of some experts, the turkey, in which it does not cause disease at any age.

Among intermediate hosts there is also some evidence of age-resistance. Thus the young of some species of snails are more readily infected with the miracidial larvae of flukes which use these snails as intermediate hosts.

At least three explanations of age-resistance have been offered. One of these suggests that the young host is not completely developed physiologically and is, for this reason, more suitable for the parasitic animal than it is when it gets older. This explanation thus regards age-resistance as a form of innate resistance. Another explanation suggests that the young host is not yet able to put forth a complete response against the parasitic animal, so that it is only partially resistant to it. This explanation classifies age-resistance as a form of naturally acquired resistance. Other explanations apply physiological facts to particular instances of age-resistance. The better resistance of older hen chickens, for instance, to the roundworm *Ascaridia galli* is ascribed to the greater production by the older bird of the intestinal mucus which protects the birds against this species. Whatever the explanation of age-resistance may be, it is affected by the factors just discussed, which affect other forms of resistance.

In addition to the age of the host, its *genetic constitution* or the *race* or stock to which it belongs may have an important effect upon its resistance to certain species of parasitic animals. Thus negroes are more resistant to infestations with the hookworm, *Ancylostoma duodenale*, than white races are; and there is evidence that they resist better than white races do the effects of the human threadworm, *Enterobius vermicularis*, the large roundworm of man, *Ascaris lumbricoides*, and the dwarf tapeworm, *Hymenolepis nana*. They are also more resistant to malarial parasites. It has, however, been found that the greater resistance of some individual negroes to the malarial parasites disappears when they are moved out of the areas in which malaria is prevalent. It is therefore probable that what seems at first sight to be a greater resistance due to a racial factor is in reality no more than a greater resistance due to repeated infections with malarial parasites from childhood onwards. Premunition, discussed above, also has to be considered. Some experts, however, insist that a racial factor does enable negroes to withstand the malarial parasites better than white races do. The genetic constitution of the host may also explain the statements made that certain breeds or strains of poultry, sheep and cattle are more resistant than others are to certain species of parasitic animals.

CHAPTER 8

SOME OTHER EFFECTS OF PARASITIC ANIMALS UPON THEIR HOSTS

We have now considered, very briefly, the tissue reactions and resistance of the host provoked by parasitic animals. There remains the second category of effects upon the host mentioned at the beginning of the preceding chapter, some of which are so important that as much space as possible must be given to them.

1. The Parasitic Animal may Exert Mechanical Effects upon the Host's Tissues

Mechanical effects may be exerted by:

(*a*) relatively large parasitic animals present either singly or in relatively small numbers;

(*b*) microscopically small parasitic animals present in large numbers.

(*a*) A good example of a relatively large parasitic animal which can, by itself, cause serious mechanical effects upon the host is the hydatid cyst (figs. 50, 67). These bladder-worms of the dog tapeworm, *Echinococcus granulosus*, are usually not large. After about 5 months of growth their diameter may be about 1 cm., but they can go on growing for as long as 20 years and they may then attain a very large size. One of the largest hydatid cysts ever found in the human body was discovered in an Australian. It contained 50 quarts of fluid. Another, found in a woman in Iceland, had a diameter of about 20 in. and contained about 3½ gallons of fluid.

At the end of the last century hydatid cysts were very often found in Icelandic people, and some experts estimated that half or a third of the people of Iceland had them. One reason for this was the fact that, at that time, something like 20,000 dogs were kept by a population of 70,000 people, and few Icelandic people then knew that the dog tapeworm is the source of these cysts. Nowadays they know more about the way in which these cysts are acquired (see Chapter 9). They therefore take precautions against infection from dogs and reduce by means of drugs the numbers of the adult tapeworms in the dogs. These and other preventive methods have been so successful that at the present time only a small fraction of the population of Iceland is infected with hydatid cysts.

The fact that hydatid cysts, even after several years of growth, do not grow much bigger than a cricket ball or an orange does not necessarily mean that they cannot cause pressure effects. Even a small cyst can, in certain situations, exert pressure on vital organs which may have serious results, and if the hydatid produces other hydatids in the manner described in Chapter 6, it is more likely that serious harm will result.

The reverse effect, namely, pressure upon the hydatid cyst by tissues of the host which are firm enough to resist the expansion of the cyst, so that its growth is restricted or its shape is altered, may also produce results serious for the host. When, for instance, hydatid cysts settle in bone, the outer layers of the cyst are poorly nourished in the dense bone tissue, and the hydatid may travel along the interior of the bone, often causing erosion of it.

All these types of effect may be caused by hydatid cysts which have only a single cavity (*unilocular hydatid*). In man and cattle in southern Europe and parts of Russia, however, a type of hydatid cyst is found which has several cavities (*multilocular hydatid*). It is believed that this kind of hydatid is formed when the growing cyst is confined in tissues which resist its enlargement, so that it breaks out of the capsule formed by the host's tissues round it and forms a number of vesicles which penetrate into the surrounding tissues. Portions of any hydatid may become detached and be carried by the blood to other organs in which they settle down and go on growing.

Another kind of bladderworm which causes serious pressure effects is *Coenurus cerebralis*, the bladderworm of the dog tapeworm, *Taenia multiceps*. This bladderworm (figs. 49, 68, 69) may settle in the brain or spinal cord of sheep and other domesticated animals which are the intermediate hosts of this species. They may grow until their diameter is about 5 cm., and their pressure upon the brain or spinal cord may cause severe symptoms. These vary according to the situation of the bladderworm in the central nervous system. The animal infected with the bladderworm may show abnormal movements, such as staggering, high-stepping and moving round in circles, or loss of balance or blindness. The name 'staggers', by which the disease caused by this bladderworm is known in some parts of Britain, refers to the disorder of the animal's movements. It is, however, a bad name for this disease, because the name 'staggers' is sometimes also given to diseases caused by bacteria and other agencies, the symptoms of which are similar. Other names given to the disease caused by *Coenurus cerebralis* are 'sturdy' and 'gid'.

The bladderworms of other species of tapeworms, although they are smaller, may also cause serious pressure effects, especially if they are numerous. Some of the symptoms of human cysticercosis, mentioned in Chapters 4 and 9, are caused by the pressure and irritation of the small, but often numerous, bladderworms of *Taenia solium* upon the human brain, eye and other organs.

Another species whose effects include pressure upon the host's tissues exerted by cysts is the fluke, *Paragonimus westermanii*, but these cysts are not bladderworms. They are not part of the parasitic animal at all, but are the result of the host's tissue reactions to the adult fluke.

Among the vital channels in the host's body which may be obstructed or completely blocked by parasitic animals are the air passages and the food canal. Some species which obstruct the air passages are discussed in Chapter 9. The food canal is perhaps less often obstructed by parasitic animals, perhaps because it is a muscular tube accustomed to propelling solid or semi-solid material along its length. When, however, large numbers of parasitic animals are intertwined in it, the resulting mass may cause intestinal obstruction. When, for instance, large numbers of the large roundworm, *Ascaris lumbricoides*, are present (fig. 62), they may block the food canal. Intestinal obstruction is especially likely to follow if drugs given to kill these parasites produce an inert mass of intertwined dead worms. It is safer, when the worms are numerous, to give smaller doses of these drugs repeatedly and thus to kill a few worms with each dose. Even individual members of this species may, when they are large, get into and block up vital channels, such as the ducts which take the bile from the gall bladder or from the liver.

Another example of the effects of obstruction of vital channels is the obstruction of the lymphatic glands and the lymphatic ducts caused by the adults of Bancroft's filarial worm. In this instance it is the flow of lymph which is obstructed, and the results which follow the obstruction caused by the roundworms themselves are increased by the tissue reaction around them, which causes the formation of much connective tissue. This connective tissue thickens the parts affected and may contract upon and constrict the lymph channels. There may also be an infection by bacteria. The net result of the obstruction to the flow of lymph caused in these ways is that lymph is dammed up and accumulates behind the obstructions. It then exudes through the walls of the lymph channels into the surrounding tissues, so that these tissues become filled with fluid and the condition known as *oedema* is set up. The parts of the host's body thus affected gradually enlarge and their tissues become thickened, and this enlargement and thickening may go on increasing slowly for several years. This is the condition called *elephantiasis*. Other results, which are not so distressingly visible, may follow the blocking of the lymphatic vessels, but they may be just as serious or even more so. The obstruction of the air passages of birds by the 'gapes' worm and the production of acarine disease by obstruction of the breathing tubes of the bees are discussed in Chapter 9.

(*b*) The presence of relatively very small parasitic animals in large numbers may block or obstruct the finer channels in the host's body, especially the capillary blood vessels and lymphatic vessels. Blocking of the blood capillaries may also be caused indirectly by the parasitic animal. Red blood cells infected with malarial parasites, for instance, show a tendency to stick together and to form clumps, so that, if these parasites are numerous, the clumping of the red blood cells may be sufficient to block the capillary blood vessels in which it occurs. If the capillary blood vessels of the brain are blocked in this way, no blood reaches the parts of the brain which these capillaries supply with blood; or the damming up of the blood by the clumped red cells may cause rupture of the capillaries, so that blood escapes into the tissues of the brain. Loss of consciousness, and symptoms resembling those of a 'stroke', which is the popular name for the effects of bursting or blocking of a blood vessel of the brain, then follow and many deaths have been caused in this way by the malarial parasites. The species most likely to cause this 'cerebral malaria' is *Plasmodium falciparum*, but occasionally *P. vivax* obstructs capillary vessels in a similar way. The blood vessels of the brain of the chicken may be obstructed by the early phases of the malarial parasite of the fowl, *P. gallinaceum*, and chickens may be killed in this manner. These are the phases which correspond to the phases of human malarial parasites which are found in the liver (exo-erythrocytic phases). *Trypanosoma evansi*, the cause of the disease of horses called surra, may be so numerous in the fluid part of the blood (*plasma*) that drugs given to kill them must, like drugs given to kill *Ascaris lumbricoides*, be given carefully in case they kill too many trypanosomes at once and cause obstruction of the blood vessels with dead trypanosomes.

2. INTRODUCTION INTO THE HOST OF OTHER KINDS OF PARASITIC ANIMALS OR OF BACTERIA OR VIRUSES

(a) Introduction of other kinds of Parasitic Animals

Parasitic animals may be introduced into the host by other parasitic animals which are either:

(i) *mechanical vectors*, in which the parasitic animal introduced does not undergo any part of its life history. This method of introduction into the host is called *mechanical transmission*;

(ii) intermediate or definitive hosts, in which the parasitic animal introduced must undergo part of its life history. This method of introduction into the host is called *cyclical transmission*.

(i) *Mechanical transmission*. There are some species of parasitic Protozoa which depend entirely upon mechanical transmission to their hosts by temporary blood-sucking parasitic animals. The mechanical transmission of trypanosomes is especially important. Some species of trypanosomes, in fact, are always transmitted by this method and never undergo any part of their life history inside the bodies of their blood-sucking vectors. Thus *Trypanosoma evansi*, the cause of the disease called surra of horses and their relatives, and of dogs and camels, and its relative, *T. equinum*, the cause of the disease called mal de caderas of South American horses, and also other species of trypanosomes related to these two species, rely upon mechanical transmission to their warm-blooded hosts by tabanid flies and stable-flies. Transmission in this manner by tabanid flies is especially likely to happen when the feeding of the fly is interrupted by the efforts of the warm-blooded host to get rid of the fly. Tabanid flies (fig. 9) are voracious and persistent feeders, and their relatively coarse mouthparts (fig. 60) cause irritating and relatively severe wounds, so that the host reacts vigorously against them. The fly's mouthparts may then be withdrawn with living trypanosomes adhering to it, and before these trypanosomes can be killed by drying, the fly may bite another host and thus transfer the trypanosomes to this new host. *T. evansi* may also be mechanically transmitted by the stable-fly, *Stomoxys calcitrans*, by certain species of ticks and possibly also by blood-sucking bats. It is also known that the two species of trypanosomes which cause human sleeping sickness in Africa, *Trypanosoma gambiense* and *T. rhodesiense*, may be mechanically transmitted in a similar manner by tsetse flies (fig. 8) which take up human blood from one human being and then bite another human being immediately afterwards. If, for instance, a tsetse fly sucks blood containing *T. gambiense* from one of a crowd of African natives, trypanosomes of this species may adhere to its proboscis, after the meal of blood has been taken, especially if that meal has been interrupted, and may then be transferred to another human being bitten immediately afterwards by the same tsetse fly.

(ii) *Cyclical transmission.* This is the normal method by which intermediate or definitive hosts introduce animals parasitic in them through the skin or outer layers of the corresponding definitive or intermediate hosts. Examples of it given elsewhere in this book are the cyclical transmission of the malarial parasites and filarial roundworms of man by mosquitoes, of the species of *Babesia* by ticks and of some of the species of *Trypanosoma* and of *Cryptobia* (*Trypanoplasma*) parasitic in fishes, by leeches. It is important to distinguish between mechanical and cyclical transmission, because parasitic animals transmitted by the former method can only be transmitted by bites which almost immediately succeed one another, while species transmitted cyclically cannot be transmitted until sufficient time has elapsed to enable them to undergo the part of their life history in the vector which must be completed before the infective phase is reached. Mosquitoes infected with the malarial parasites of man cannot, for instance, transmit them to man until their sporozoites have been formed, and the time required for this will vary according to the species of malarial parasite, the climatic and other factors affecting the development of the parasite in the mosquito and other factors, some of which are discussed in Chapter 10. A tsetse fly infected with *Trypanosoma gambiense* cannot transmit this species of trypanosome cyclically—although it has already been explained that it may do so mechanically and therefore immediately under certain conditions—until it has undergone its development in the tsetse fly, which requires from 15 to 35 days.

The instances just given of the introduction of one kind of parasitic animal by another kind which is its intermediate or definitive host are instances of their injection into the interior of the host. But there is another way in which one blood-sucking species may introduce another kind of parasitic animal. It is exemplified by the method by which man and animals may become infected with *T. cruzi*, which causes South American trypanosomiasis (Chagas's disease). This trypanosome is transmitted by blood-sucking triatomid bugs. Two species of these inhabit human dwellings and communicate this trypanosome to man. There are, in Brazil, the large, black and red triatomid bug, *Triatoma megista*, and in Argentina, *T. infestans*. In Venezuela, a relative of these bugs, *Rhodnius prolixus*, also infects man. Some twenty other species of triatomid bugs have been found infected with this species of trypanosome, but many of them do not normally bite man.

The trypanosome undergoes a development in the food canal of the bug which produces infective phases that pass out in the bug's droppings. If these infected droppings are present on the host's skin when the host rubs or scratches the irritated wound produced by the bite of the bug, the infected droppings may be rubbed into this wound, so that the host infects itself with the trypanosomes. Rats may infect themselves with *Trypanosoma lewisi* by rubbing the droppings of the rat fleas, which are the intermediate hosts of this trypanosome, into the irritating bites of the fleas. *T. cruzi* can also enter man when infected droppings of the triatomid bug come into contact with the membrane lining the eyelids (conjunctiva) or the mucous membrane of the mouth, or if the bugs are accidentally swallowed; and rats may infect themselves with *T. lewisi* by swallowing infected rat-fleas (cf. Chapter 9).

(b) *Introduction of Bacteria and Viruses*

The introduction of bacteria by the biting mouth parts of mosquitoes, midges and other blood-sucking species is well known. Severe and even fatal results have followed when certain bacteria have thus been injected into man.

More frequently a temporarily parasitic animal, such as a mosquito, or a permanently parasitic species, such as one of the mites which cause scabies of man and animals, creates irritation, and the host rubs or scratches the irritated area and at the same time introduces bacteria into the irritated tissues. In this way septic sores may be caused. They are a common result of the irritation caused by the mites which cause scabies and may result even when the irritation is set up by animals, such as the biting lice of poultry, which do not themselves injure or abrade the skin of the host upon which they live.

Secondary bacterial infections of internal organs may also occur when these are penetrated by adult parasitic animals or their larvae. It readily happens when the walls of the food canal are penetrated or injured by parasitic animals, because the contents of the food canal are full of bacteria. The injuries inflicted on the large intestine of man and animals by *Entamoeba histolytica* (fig. 17), for instance, which cause human amoebic dysentery, are usually supplemented by the effects of bacteria which may do more harm than the Entamoebae do. The injuries inflicted by roundworms living in the host's food canal are also often infected by bacteria derived from the contents of this canal. When the larvae of these roundworms burrow into the intestinal wall and undergo part of their development there, as the larvae of the nodular worms (*Oesophagostomum*), for instance, do, bacteria readily infect the injuries which these larvae leave behind themselves. When the eggs of the human blood flukes irritate the walls of the bladder or rectum of man and make their way through these walls, the areas which they have injured often become infected with bacteria.

The instances so far given of bacterial infection of injuries caused by parasitic animals are all instances of secondary bacterial infection, that is to say, of the entry of bacteria into tissues already injured by the parasitic animal. But there is some evidence that adult parasitic animals, or their larvae, can be themselves infected with bacteria, or with viruses, before they enter the host's tissues and damage them. When this is so, they actually transport the bacteria or viruses into the host's body. It has been stated, for instance, that anaerobic bacteria or their spores can be introduced in this manner into the livers of hosts of the liver fluke, *Fasciola hepatica*, and that bacteria can be inoculated into the walls of the intestine by the heads of tapeworms which attach themselves to these walls.

When these bacteria have been taken into the intestines of infective larval phases of parasitic animals, it seems probable that these infective larvae can carry them into the tissues of the host into which they penetrate, so that they introduce the bacteria and viruses just as a hypodermic needle containing them would do. It is difficult to prove experimentally that this actually does happen; but French workers have claimed that they have proved experimentally that the bacilli which cause tuberculosis and anthrax can be taken into the host's body by the infective larvae of hook-worms, when these larvae burrow through the skin, and that these bacilli can kill guinea-pigs into which they are carried in this way. Australian work has indicated that the bacillus which causes foot-rot of sheep may be assisted into the feet of these animals by the penetration of the membrane between the sheep's toes by infective larvae of the roundworm *Strongyloides papillosus*. There is some evidence, too, that the virus of pig influenza can be introduced into healthy pigs by the larvae of lungworms of the pig, or even by the earthworms which are the intermediate hosts of these larvae. It is also claimed that the larvae of *Trichinella spiralis* can introduce into guinea pigs the virus which causes one kind of inflammation of the membranes enclosing the brain (*meningitis*).

3. Production by the Parasitic Animal of Substances Injurious to the Host

These may be either true toxins, which are substances produced by the normal physiological processes of the parasitic animal, or substances set free into the host by the death and disintegration of the parasitic animal.

(a) *True toxins*. A toxin is a substance which, when it is injected by itself into an animal, produces harmful effects. So far as we know at present, only one substance, produced by a single genus of parasitic animals, fulfils this definition. This is the substance called *sarcocystin*, which has been extracted from the protozoon *Sarcocystis tenella*, which infects sheep. Sarcocystin will kill experimental animals.

(b) Substances produced by the normal physiological processes of parasitic animals may be *secretions*, which are useful to the parasitic animal, or *excretions*, which it does not require. Both may be set free into the host's tissues and both may act, either locally on the tissues surrounding the parasitic animal, or, when they circulate in the host's blood and tissue fluids, on other tissues and organs situated elsewhere.

Examples of substances produced by parasitic animals which have a localised action on the host's tissues are the digestive ferments secreted by whipworms which convert portions of the walls of the host's food canal into fluid upon which these roundworms feed; the anticoagulins of the hookworms and leeches which prevent the clotting of the host's blood; the substances injected into the host by mosquitoes, stable-flies, lice, ticks, bugs and other temporarily or permanently ectoparasitic species and the substances (cytolysins) secreted by such protozoan species as

Entamoeba histolytica which dissolve the walls of the human large bowel, so that ulcers are produced in it, into which bacteria may enter and aggravate the effects of the Protozoa. The ciliate protozoon, *Balantidium coli* (fig. 21), which is parasitic in the large intestine of man and pigs, probably feeds directly on the tissues and red blood cells of its hosts.

There is, however, at least one group of parasitic animals which, when they become parasitic in suppurating wounds, may have beneficial rather than deleterious effects upon these wounds. These are the larvae of certain kinds of blowflies (see Chapter 9), from whose secretions bactericidal substances have been extracted. These maggots may get into septic human wounds, and as long ago as the sixteenth century it was noticed that wounds infected with them healed more quickly. Not only do they kill bacteria in the wounds, but they also eat up dead tissue and, by making the wounds alkaline, they help their healing.

There are plenty of examples of substances passed out of the bodies of parasitic animals into the host which take effect beyond the part of the host's body into which they are passed. They may produce characteristic changes in the relative numbers of the different kinds of blood cells and corresponding changes in the organs from which these cells are derived. There may be, for instance, enlargement of the spleen, a change which may be marked in people, especially children, who suffer from chronic malaria. Enlargement of the liver, lymphatic glands and bone marrow may also be signs of this disease. These enlargements are the expression of an attempt made by the host to combat the extensive destruction of red blood cells by the malarial parasites.

Many other kinds of parasitic animals pass into the host substances which cause similar changes in organs remote from the sites occupied by the parasitic animals. It is not possible to describe these here. Some of them are the expression of unusual activities of the organs concerned; some of them are definitely associated with immunological activities. The spleen undoubtedly plays a part, which is not yet fully understood, in the establishment and maintenance of resistance to infections with parasitic animals. When it is surgically removed, some kinds of parasitic animals establish themselves more readily in these splenectomised hosts. The tissues of the spleen are themselves often invaded by some kinds of parasitic animals, which may lie up in this organ and form a reservoir of parasites which periodically multiply and renew their characteristic injurious effects. The liver and the lymphatic glands are also important factors in the battle of the host with various kinds of parasitic animals.

Another class of substances which may pass into the host's body from the parasitic animal is made up of substances set free when parasitic animals die inside the host. These may be especially important when the dead parasitic animals are relatively large or numerous. When, for instance, flukes or roundworms die in their hosts, substances derived from their dead bodies may act as poisons to whose action some of the symptoms of the infection are due; or they may act as antigens which provoke a resistance, if this has not already been produced; or they may sensitise the host in such a way that it shows symptoms similar to those of asthma and other allergic conditions.

The possibility that the host may have been immunised in this way must be remembered when serological tests

are done to detect infection with parasitic animals, because, if the host has been thus immunised by an infection
that is no longer active, this earlier infection may have an
important bearing upon the interpretation of the results
of serological tests. Tests for the presence of parasitic
animal *A* may, that is to say, be affected by an earlier, but
no longer active, infection with parasitic animal *B*, which
has caused the production of antibodies that react with
antigens derived from both *A* and *B*.

Serological tests may also be affected by the sensiti-
sation of the host by substances derived from the bodies
of parasitic animals which have died in it. The introduction of even minute amounts of antigens derived from
parasitic animals to which the host is sensitised may pre-
cipitate serious anaphylactic reactions in the host, which
may sometimes kill it. If, that is to say, a host has become
sensitised to parasitic animal *A*, it may suffer anaphylactic
shock which may kill it either (*a*) when a living individual
of parasitic animal *A* dies in it and liberates the specific
antigens contained in its body; this may happen, for example, when a cow suffering from the lesions caused by
warble-fly grubs seeks to allay the irritation of these lesions
by rubbing against a fence or tree and ruptures a warble-fly
grub, so that the tissue fluids of the grub are released into
the body of the cow; it may also happen when a hydatid
cyst is ruptured in its host; or (*b*) when man seeks, by means
of serological tests, to detect the presence in the host of
parasitic animal *A* and, in order to do these tests, artificially introduces into the host small amounts of antigens
which he has prepared from the dead bodies of parasitic
animal *A*.

4. THE PARASITIC ANIMAL MAY REDUCE THE HOST'S RESISTANCE TO OTHER SPECIES OF PARASITIC ANIMALS OR TO BACTERIA OR TO VIRUSES

This important effect of the parasitic animal, and its reverse, the reduction by the effects of bacteria or viruses of the resistance of the host to parasitic animals, have been mentioned elsewhere in this book. The principle applies, not only to the host as a whole organism, but also to its individual tissues and organs. A liver injured by the migratory larvae of *Ascaris lumbricoides*, or by young liver flukes wandering through it, will offer less resistance to other parasitic animals or to bacteria or viruses than a healthy liver will. The spores of the bacterium *Clostridium oedematiens*, for example, lie dormant in the livers of many healthy sheep; but if these livers are damaged by young liver flukes wandering through them, the spores of the bacterium are enabled to multiply and to cause the disease called 'black disease', which afflicts sheep in northern England, northern Scotland, Devonshire, Shropshire and also in Australia.

Conversely, injury inflicted by bacteria or viruses may reduce the resistance of the bowel wall to infections with Coccidia or roundworms, and a lung infected with bacteria will be more vulnerable to the effects of lungworms or flukes. Even the artificial immunisation of the host to certain kinds of bacteria or viruses may so affect the host's health that parasitic animals already present in its body may be enabled to multiply and cause disease. Thus immunisation of cattle against rinderpest may light up latent infection with *Babesia bigemina*, species of *Theileria*, trypanosomes or Coccidia, and these parasitic animals may then cause the deaths of the immunised cattle. Some bacteria may alter tissues which they infect so much that these tissues become media favourable to the growth of parasitic animals. Thus bacteria provide, for the maggots which cause myiasis (see Chapter 9), the pus on which they thrive; they provide it also for the larvae of the warble-flies. It would be interesting to discuss further this kind of co-operation between different kinds of parasitic animals, but it cannot be further considered here.

5. The Parasitic Animal may Consume the Host's Food or Important Ingredients of it

Consumption of the host's food by parasitic animals may occur, not only in its food canal, but also elsewhere in the host's body. For some species of parasitic animals take food materials from the host's blood and other tissue fluids while these fluids are transporting food elements to various organs. Other species take food from stores of certain kinds of food held by the host in certain organs. They may deplete, for instance, the stores of glycogen in the liver and the muscles; or fat laid up in various parts of the body. Intracellular species, of course, take food from the interior of cells of the host.

Parasitic animals which steal ingredients of the host's food while it is still in the host's food canal include single-celled Protozoa, among which there are many amoeboid, flagellated and ciliated species, and many species of round-worms, tapeworms and flukes.

The effects of all these different species upon the food of the host cannot be described here in detail. Even if space were available for such a discussion, it would be largely speculative, because we have, as yet, comparatively little exact knowledge of the nutrition of species parasitic in the alimentary canal and less of their effects upon the nutrition of the host. Although many of these species seem to extract only trivial amounts of the host's food, others seem to deprive the host of essential nutritive substances. Recent work has suggested, for instance, that some species of tapeworms not only absorb carbohydrate from the contents of the intestines of their hosts, but also absorb from the mucous membrane which lines the host's food canal nitrogenous substances, and perhaps also vitamins belonging to the vitamin B complex. It has been suggested, therefore, that certain symptoms of tapeworm infestations, such as nervous symptoms, loss of weight and general weakness, may be due to the loss of these vitamins by the host.

It has also been suggested that certain other symptoms, such as the anaemia which may be caused by such species as *Diphyllobothrium latum*, are produced by the absorption by this species of tapeworm of the so-called 'intrinsic factor' produced by the host, which converts an 'extrinsic factor' contained in meat into a factor which prevents the occurrence of anaemia. This view contrasts with the hypothesis which postulates that the anaemia caused by *Diphyllobothrium* is due to the absorption by the host of poisonous substances produced by the tapeworm. Although some species of tapeworms, and especially the larger species, do thus affect their hosts by appropriating certain elements of their food, some of the smaller species, such as the species of the genera *Davainea* and *Raillietina*, which infest poultry and other birds, seem to affect their hosts principally by causing inflammation, ulceration and bleeding. This kind of damage is the more serious because these species are often present in large numbers.

Another species that damages the walls of its host's intestine is *Hymenolepis nana*, described in Chapter 4, which is parasitic in man; but in this instance, the damage is done by the larvae of the tapeworm, while they develop in the intestinal wall. Human children especially may harbour enormous numbers of this species. In parts of Russia, for example, it is not unusual to remove, by means of drugs, two or three thousand specimens of *H. nana* and Russian authors have reported the removal of tens and even hundreds of thousands of them from children severely starved and debilitated during their detention by the Nazis during the recent World War. When so many tapeworms are present, they can hardly fail to affect the host both by injury of its intestinal walls and by absorption of certain elements of its food. Hosts which are poorly fed and therefore less resistant will suffer more severely from both these kinds of effects.

6. The Parasitic Animal may Consume other Substances which are not the Host's Food, but are Essential to its Life or Good Health

The study of this question opens up a very wide field of inquiry, but discussion of it is restricted by our lack of adequate knowledge of the physiology of parasitic animals.

A very important example of substances which are not food which may be taken from the host by parasitic animals is the red pigment, haemoglobin, which gives the red colour to the red cells of the blood. This pigment carries the oxygen taken in by the breathing organs to the various tissues of the body, so that the host cannot do without it. When a certain amount of it is lost, or when it is being lost so rapidly that the organs of the body which make it cannot make it quickly enough to replace the losses of it which are occurring, the animal begins to suffer from anaemia; and to this anaemia can be traced a series of effects which undermine the host's health and reduce its resistance to the effects, not only of parasitic animals, but also of other causes of disease. A host which is rendered anaemic by one species of parasitic animal may therefore succumb to the effects of another species of parasitic animal or to those of bacteria or viruses.

Parasitic animals which may thus cause anaemia may be (a) species which live only on the surfaces of their hosts (ectoparasites) or (b) species which live inside the bodies of their hosts (endoparasites).

(a) Blood-sucking Species Parasitic on the Host's Surface

Among these are the blood-sucking bats considered in Chapter 5 and the leeches and ticks.

It is difficult, for more than one reason, to estimate the amount of blood removed by ticks. One reason is that the adult tick, like the adult blood-sucking roundworm, passes out in its droppings much of the blood taken in. Another is the fact that it is very difficult to estimate the amounts of blood taken in by the larvae and nymphs as well as that taken in by the adult ticks. We have also to consider whether the tick under consideration is a one-, two- or three-host tick (see Chapter 9). The larva, nymph and adult of the three-host tick feed on different individual hosts; the larva and nymph of a two-host tick feed on one individual host and the adult on another; while the larva, nymph and adult of a three-host tick each feed on a different individual host. Estimates of the amounts of blood removed by ticks must, therefore, be largely provisional. It has, however, been estimated that 1000 larvae of *Ixodes ricinus* could remove about 5 c.c. (about a teaspoonful) of blood. We do not know how much the nymphs of this species take in, but the adult female *I. ricinus* takes in about 1 c.c., and it has been estimated that a heavily infested sheep may lose about 250 c.c. (about half a pint) of blood a week. Some experts think that the loss of blood extracted by ticks does sheep little harm, while others think that it may cause anaemia.

Among the leeches there are species which cause, if they are sufficiently numerous on the host, appreciable anaemia. The small land leech, *Haemadipsa zeylanica*, which lives on the surfaces of trees and grass, or under stones and other solid objects, in the damp, tropical areas of parts of India, the Philippine Islands, Australia and South America (the Chilian Andes), attacks the uncovered skin of man and animals and can bite through clothing. It is 2·3 cm. long and may lurk in an erect position at the edges of footpaths ready to attack men and animals passing by. In Ceylon it is a serious pest. Groups of these leeches may be seen hanging from the ankles of natives or from the fetlocks of horses. Severe anaemia and irritation may be caused by them and the deaths of men and animals have been attributed to them, but their effects are not usually fatal.

A species of leech which is a troublesome pest of cattle, sheep, pigs, dogs and other animals, including man, is *Limnatis nilotica*, a large leech which is 8–12 cm. long. It is found in North Africa, mid-Europe and the Near East. It is said to be a troublesome pest of cattle in Bulgaria, and French troops in Egypt years ago suffered from its attacks.

It may be swallowed with the drinking water and may then attach itself to the cavities of the nose, or to the pharynx, tonsils or vocal cords which produce the voice in the voice box (larynx). In these situations it sucks blood and may survive for a considerable time. Man may also swallow the young leech of this species and suffer from its effects. It causes swelling of the tissues to which it is attached, and the consequences of this swelling are headache, hoarseness of the voice, cough, exudation of blood-stained froth, difficulty of breathing and even fatal prevention of the intake of air (*asphyxia*). If these leeches are numerous, they may cause severe anaemia, and if their bites become infected with bacteria, the results are correspondingly more severe. A relative of this species, *L. africana*, attacks man, monkeys, and dogs in Senegal, the Congo, India and Singapore. When people are bathing, this species may enter the vagina or urethra and cause bleeding from these parts of the body. A Japanese species of this genus, *L. japonica*, has been found attached to the membrane which lines the socket of the eye (*conjunctiva*).

(*b*) *Blood-sucking Species Parasitic inside the Host's Body*

Blood-sucking species of this group usually cause a more severe anaemia than ectoparasitic species do. The latter are less numerous and are parasitic only temporarily, while they suck blood, whereas the internal species are more numerous in any individual host and are permanently parasitic and continue to suck blood for longer periods of time.

Among the most injurious of them are the hookworms. These roundworms, which have a formidable sucking apparatus and efficient teeth (fig. 56), produce, and inject into the host, anticoagulins which prevent the clotting of

the host's blood and thus ensure for the hookworm a continuous supply of it. Some species of them, moreover, suck
arterial blood only and appear to select small arterioles
rather than small veins. They also have a habit of causing
bleeding at one site and, after a period of blood-sucking at
this site, moving on to other places, leaving behind them
damaged blood vessels which go on bleeding after the hookworms have left them. The loss of blood from these bleeding
points may add considerably to the anaemia caused.

The amount of blood lost by a host infected by hookworms may be astonishingly large. American and Japanese
workers, who have watched the dog hookworm, *Ancylostoma caninum*, sucking blood from its host's small intestine,
have calculated that a single hookworm of this species can
remove at least $\frac{1}{2}$ c.c. of blood each day. Fifty of these
hookworms could therefore remove nearly two tablespoonfuls daily, and five hundred of them could cause a daily loss
of 250 c.c. (nearly half a pint) of blood. If the bleeding
from sites abandoned by the hookworms is included in the
estimate, the total loss of blood is greater than this.

The amount of blood removed by human hookworms is,
of course, not necessarily the same as this, but the human
and dog hookworms are close relatives, and there is no
reason to suppose that their relative capacities for sucking
blood differ appreciably. If it is assumed that the human
hookworms do remove as much blood as the dog hookworms do, namely, 250 c.c. a day, and, if we reckon, as
experts do, that the total volume of the blood of a healthy,
average human being is about 6 litres, which is about
$10\frac{1}{2}$ pints, five hundred hookworms could remove every day
about one twenty-fourth of the total volume of the blood.
The replacement of the haemoglobin and also of the fluid

content of the blood represented by this degree of loss must be rapid if the host's health is to be maintained.

But, the reader will ask, do human beings suffer from infections with as many as five hundred hookworms? The answer to this question is that they may harbour many more. It is difficult to estimate exactly the number of hookworms present in any particular person, but individuals have, after single doses of suitable drugs, expelled more than a thousand of them. It is probable that the rate of blood-sucking stated above falls during the life of each hookworm to at least a fifth of the rate stated and the resistance of the host (see Chapter 7) comes into action to help the remarkable capacity of the host's blood-forming organs to replace losses of blood. At all events, at least 500 hookworms are needed to cause a state of disease in a healthy adult taking normal amounts of the iron necessary for the manufacture of haemoglobin. Light infestations are usually harmless to healthy adults, but even 25 hookworms may cause disease in poorly nourished individuals. Children and pregnant women (whose iron is being depleted by the demands of the young in their wombs) may suffer severely. In children 100 hookworms may cause disease.

The reason why hookworms remove these prodigious quantities of arterial blood is not known. It seems unlikely that they require the oxygen in arterial blood, because our scanty knowledge of the physiology of roundworms suggests that they need only very small quantities of oxygen, if, indeed, they require any at all. It is known that some species of roundworms make, as some other kinds of animals also do, varieties of haemoglobin which are peculiar to themselves, and it is possible that hookworms suck the blood of their hosts because they can make out of its haemoglobin their own particular variety of this substance; but if they do this, we do not know what use they make of the haemoglobin manufactured. It is also possible that the phosphorus in the host's blood is required by the roundworms for the large numbers of eggs that they produce. The amount of phosphorus in the eggs of the sheep stomachworm, *Haemonchus contortus*, has, indeed, been used to estimate the amount of blood removed from the abomasum of sheep by this species. It was estimated by this method that 4000 stomachworms, which is a reasonable average of the number often found in sheep, would remove about 60 c.c. (about 4 tablespoonfuls) of blood each day.

Apart from the blood and the food of the host, other substances in its body may be consumed by parasitic animals, and some of these have been mentioned earlier in this book. It would be interesting to pursue this subject further, but limitations of space unfortunately prevent us from doing this.

7. The Parasitic Animal may cause Biological Effects

Apart from their effects upon the tissue and tissue fluids of the host, some parasitic animals cause profound alterations in the physiological processes of the host which may be termed *biological effects*. Remarkable examples of these are the profound changes caused by species that cause the changes in the host's reproductive glands to which the name *parasitic castration* is given. The word castration is, however, not strictly appropriate to these changes, because it means destruction or removal of the male genital gland only, whereas the effects about to be described include injury or destruction of the female genital gland and also sterility due to injury to either the male or the female glands or to both.

A species which causes effects of this kind is the parasitic crustacean *Sacculina* (fig. 58), whose structure is profoundly modified by parasitic life in the manner described in Chapter 5. This genus belongs to the family Rhizocephalidae, all the members of which are parasitic, and this family belongs to the order Cirripedia, of which the barnacle is an example.

The effects of *Sacculina* upon its host are remarkable and severe. They are well illustrated by *S. neglecta* and its host, the short-tailed spider-crab, *Inachus mauritanicus*. First, the growth of some hosts may be hindered or prevented because the presence of *Sacculina* delays or prevents the shedding of the skin and exoskeleton upon which the growth partly depends. The growth of other hosts of *Sacculina* is, however, accelerated, and this quickening of the growth is associated with the destruction by *Sacculina* of both the male and the female reproductive organs. The effect is therefore similar to that which occurs when male chickens are castrated to produce capons, which fatten up more quickly than normal chickens do.

In addition to this effect, about 70 % of male crabs acquire, as a result of the effects of *Sacculina* upon them, some of the secondary sexual characters of the female. The abdomen of these males becomes broad, they may acquire, in addition to their male copulating styles, appendages modified to bear eggs and their nippers (*chelae*) become smaller at the same time.

Sacculina, however, does not cause female crabs upon which it is parasitic to acquire secondary sexual characteristics of the male. Their egg-bearing appendages may become smaller, but they do not acquire male external characteristics.

The reproductive organs of both the male and female crabs parasitised by *Sacculina* show varying degrees of disintegration and degeneration. When, however, the parasitic *Sacculina* drops off, the parasitised male which has acquired secondary sexual features of the female may then regenerate, from the remains of its reproductive organs, not a testis, but a hermaphrodite sexual gland, so that the males then produce both eggs and sperms. If, however, the *Sacculina* drops off the female crab, she cannot regenerate the ovary. Another species which produces similar effects is *Peltogaster curvatus*, which is parasitic on the short-tailed hermit crab, *Eupagurus excavatus* var. *meticulosa*.

Even more remarkable than these effects produced by *Sacculina* upon the genital organs of its hosts are those which some parasitic isopod Crustacea can produce without any contact with these organs. The order Isopoda, to which the wood-lice or sow-bugs belong, includes species which are parasitic upon other Crustacea or even upon other parasitic Isopoda and upon the parasitic Rhizocephalidae, belonging to the genera *Peltogaster* and *Sacculina*, which have just been mentioned.

Other species of parasitic Isopoda, which belong to the family Entoniscidae, are at first ectoparasitic in the gill-chambers of their crab hosts and later invaginate the wall of this chamber into the body cavity of the host, so that they are still ectoparasitic, although they now feed upon the fluids in the host's body cavity, which they absorb through the wall of its branchial chamber. Species of the genus *Entoniscus* are parasitic upon British shore crabs, belonging to the genus *Porcellana*, and *Portunion maenadis* is parasitic upon the common British shore crab, *Carcinus maenas*. When a crab is infected by *Sacculina* or *Peltogaster*, it is more likely to become infected by parasitic Entoniscidae. It has been suggested that, because the former parasite prevents or delays the moulting of the crab's exoskeleton, the entoniscid is exposed to fewer risks during its life.

Some species of hymenopterous insects also injure the reproductive organs of both sexes of their hosts. Species of the family Dryininae, for example, whose larvae only are parasitic upon the nymphs of various Homoptera (leaf hoppers, tree hoppers, cuckoo-spit insects, etc.), cause degeneration or disappearance of the genital organs of both sexes of the host, and the abdomen of the male host may acquire the pigmentation, shape and texture of that of the female of its species. Thus species of the genus *Aphelopus* may produce these effects upon members of the genus *Typhlocyba*, one of the jassid leaf hoppers, and upon membracid tree hoppers belonging to the genera *Thelia* and *Telamona*. The hosts of the Dryininae may react to their effects by proliferation of the integument, so that an external cyst is formed, which resembles the galls which grow upon some plants. One or several of these cysts may be found upon various parts of the body of one host and they may be as large as the host's abdomen. Usually the cysts are yellow or black. The Dryininae leave the host to pupate on the soil or on the host's food plant. The larvae of the two-winged dipterous Pipunculidae, parasitic upon homopterous insects belonging to the genera *Chalaris* and *Typhlocyba*, may produce similar 'castration' effects.

Just as remarkable as the effects of *Sacculina* and its relatives upon crabs are the effects of insects belonging to the order *Strepsiptera* upon their hosts. These insects have been given the name *Stylops*, so that the effect of these insects is sometimes called *stylopisation*. The hosts of Strepsiptera include hymenopterous insects, among which are wasps belonging to the genera *Vespa* and *Polistes*, solitary bees belonging to the genera *Andrena* and *Halictus*, which live in villages, and homopterous insects, among which are various leaf hoppers.

The larva and the female strepsipteran are both parasitic, but the male is not. The male leaves the host for a brief, winged life lasting a few hours only. The female never leaves the host. She is buried in it, with only her cephalothorax protruding through the body wall, and in this position her eggs are fertilised by the male.

The eggs of some Strepsiptera appear to develop parthenogenetically. The larvae hatch in the body of the female. They pass out of her, and reach the surface of the body of the host, on which they are very active. They probably reach other bees and wasps by getting on to flowers and from these on to bees and wasps visiting the flowers after them. By these new hosts they are transported to the nests, or they may reach the nests directly from infected hosts.

In the nests they burrow through the body wall of the larvae of the host. Inside the host the parasites grow and moult their skins and become legless 'maggots'. The larvae of the species *Xenos vesparum* feed by absorption of the host's blood through their skins. The parasite lies between the host's organs, not in them, and may push them out of position. The host's organs are not directly damaged, but may suffer from malnutrition. By the time that the wasp or bee larva has pupated, the strepsipteran larva has protruded between its abdominal segments. The male strepsipteran then pupates and eventually emerges for its brief winged life. The female remains in the wasp or bee.

As many as thirty-one strepsipteran parasitic larvae have been found in one host; but female hosts are most often attacked. Both sexes of the strepsipteran may be found on the same host, but usually males are commoner.

The effects upon the host consist chiefly of reduction of the parts affected, but the male hosts suffer less than the female hosts do. The head of both sexes of the host becomes smaller, more globular and more hairy. The pollen-collecting apparatus of the female hosts is much reduced, so that the hind legs resemble those of the males, and the yellow colour of the male may be acquired. The sting is reduced in size. The ovaries become smaller, the oocytes in them degenerate, and there is no evidence that parasitised females are ever fertile. In males the copulatory apparatus is reduced. Although the testes, like the ovaries, become smaller, they can produce sperms.

Some parasites of insects, which are not themselves insects, may cause profound effects upon the sexual development of the insect host. The roundworm family, Allantonematidae, shows us several examples of this. *Tylenchinema oscinellae*, parasitic in the body cavity of *Oscinis frit*, the frit-fly, which does great damage to oats, wheat, barley and grasses, sterilises both its male and female hosts. Probably its larva actively bores through the body wall of the larva of the host by means of the stylet which it possesses and perhaps also with the aid of secretions which it passes out. Its life history is correlated in an interesting way with the three generations per year produced by the host. Another interesting feature of this host-parasite relationship is the fact that occasionally the host does develop male or female genital organs, and when it does the roundworm fails to grow and develop.

The life histories of some roundworms which are parasitic upon insects are complicated by the inclusion of one or more parthenogenetic generations. *Heterotylenchus aberrans*, which is parasitic upon both sexes of the river-fly, *Hylemyia antiqua*, has these. The males of this fly are not, however, sterilised, although the ovaries of the females fail to develop.

The roundworm *Sphaerularia bombi* is another interesting species which causes the sterility of the queens of humble-bees and some wasps (*Vespa rufa* and *V. vulgaris*). A remarkable feature of the life history of this species is the fact that, after the entry of the female roundworm into a new hymenopterous host, the actual body of the female roundworm does not grow. Instead, its uterus is everted and prolapsed through its external genital opening, taking inside it the reproductive organs and the modified intestine. This prolapsed uterus then grows enormously, and the rest of the roundworm's body remains minute and functionless and may be cast off.

The members of the roundworm family Mermithidae produce effects similar to those which have just been described, although they do not necessarily destroy the reproductive organs of their hosts. In Chapter 3 it was explained that the larval stages only of the Mermithidae are parasitic, the adults being non-parasitic in the soil. The hosts of the larvae are ants, crickets, grasshoppers, earwigs and some other insects. Usually the development of the host's reproductive organs, and especially that of the testes, is arrested, so that the insects become sterile. The other organs of some hosts of Mermithidae are not appreciably affected, but those of some ants certainly are. The ants *Lasius alienus* and *Pheidole absurda* may be so altered by Mermithidae that they do not show the characters of either the female, the worker or the soldier. Ants modified in this way are called *intercasts*.

8. The Parasitic Animal may Seriously Affect the Life of Communities of Animals, including Human Communities

A full account of the effects of parasitic animals upon human communities would, of course, fill a book of this size. Some of them have been already implied in the preceding pages; others are indicated in Chapters 10 and 11; and there are others still which, vital though they are to the progress of human civilisation, cannot be discussed in a book of this kind. A brief note, however, must be given on the effects of some kinds of parasitic animals upon communities of ants, bees and wasps, a phenomenon to which the name *brood parasitism* is given. Large monographs have been written about this subject, and these can be consulted by readers especially interested by it. It can be only briefly considered here.

In the nests of ants a large number of species of insects, and of some other animals belonging to the large group (Arthropoda) to which the insects belong, live as guests. They are called myrmecophiles. Over two thousand different species of them have been recognised, about half of which are beetles. Some of these myrmecophilous species are predators or scavengers, to which the ants are hostile. Some of them are tolerated by the ants, although they steal the food of their hosts, and may be modified so that they

closely resemble them. Yet other species are fostered and
even reared by the ants, because they secrete substances
which the ants find attractive. The larvae of some species
of these guests are, however, parasitic upon the ant com-
munity in the sense that they eat many of the eggs and
young of the ants. They may be compared with the Mermithidae and the truly parasitic hymenopterous species mentioned in the preceding section.

In addition to these kinds of association between ants and
other species of animals, there are the various grades of
what is called *social symbiosis* between different species of
ants. These are connected by intermediary gradations with
the practice of slave-making and with what is called
temporary social parasitism, which is exemplified by those
queens that enter the colony of another species, which
always belongs to the same genus or to one which is closely
related. The first brood of the young of the invading queen
is then reared by the workers of the invaded colony. This
kind of behaviour also occurs among the bees, the queens of
some species of which may enter another colony and kill the
queen of that colony.

Thus the queens of colonies of *Bombus leucorum* may be
killed in this way by invading queens of the species *B. terrestris*, the workers of which species rear the young of the
B. terrestris queen. A more specialised parasitic habit is
shown by bees belonging to the genus *Psithyrus*, which
produce no workers, nor has the queen any pollen-gathering
organs on her hind legs. She enters the nest of species of
Bombus in the spring and 'ingratiates herself' to the workers
in it, so that they soon cease to be hostile. She then kills the
Bombus queen and the *Bombus* workers rear her larvae. The
species of *Bombus* and *Psithyrus* are very much alike, and

some experts believe that the latter genus arose from the former. A similar kind of behaviour is exhibited by some species of wasps.

The effects of parasitic animals upon their intermediate hosts

A note on this interesting subject is included here to remind the reader that we have already considered some serious diseases of man and his domesticated animals which are caused by phases of parasitic animals, such as the malarial parasites and some kinds of tapeworm, of which man and his domesticated animals are intermediate hosts.

Man is also among the intermediate hosts of the plerocercoid larvae of some species of tapeworm related to the human fish tapeworm, *Diphyllobothrium latum* (see Chapter 4). The adult phases of these larvae are not all known, but they develop, like the plerocercoids of the fish tapeworm, from procercoid larvae parasitic in *Cyclops*. They may be found in the muscles of frogs, snakes and some small mammals, as well as in human tissues, and, like all plerocercoid larvae, they can pass from one host to another until they are eaten by a definitive host in which they can become the adult tapeworm. When they were first found in man, their nature was not understood and they were given the name *Sparganum*, the disease that they cause being called *sparganosis*. Man may acquire these plerocercoid larvae in the Far East and in parts of Africa and Holland, by swallowing *Cyclops* containing the procercoid larvae or by migration of the plerocercoid larvae directly into human tissues when the uncooked flesh of hosts containing them comes into contact with the human body. In the Far East, for example, the uncooked flesh of frogs is applied as a poultice to inflamed eyes or wounds and plerocercoid larvae (Spargana) in it may then directly penetrate into the human tissues thus treated. In these tissues some kinds of these plerocercoids branch and multiply. They cause inflammation, swelling or a form of elephantiasis. The plerocercoid larvae of related species of tapeworms, such as *Ligula* and *Schistocephalus*, are found in the body cavities of the stickleback and other fish. They are the plerocercoid larvae of adult tapeworms which are parasitic in the food canals of aquatic fish-eating birds. They do not normally infect man and seem to do little harm to the fish.

Many other invertebrate intermediate hosts appear to be comparatively well adapted to the larvae of parasitic animals which live in them. Thus Cyclops apparently suffers little from the effects of either the procercoid larvae of tapeworms or the larvae of the guinea-worm. Snails which are the intermediate hosts of the nematode lung-worms described in Chapter 9 may harbour as many as 200 of the larvae of these worms without suffering any apparent harm. On the other hand, the miracidial larvae of flukes may irritate the surface tissues of the snails into which they enter. The rediae of *Fasciola hepatica* may kill *Limnaea truncatula* and the sporocysts of *Schistosoma haematobium* form secondary sporocysts which ramify in the digestive gland (*hepatopancreas*) of its snail intermediate host and may almost completely destroy that gland. The miracidia of *S. mansoni* penetrate within a period of a quarter of an hour the tentacles of the species of snail belonging to the genus *Planorbis*, which are among their intermediate hosts and cause at their point of entry deformation and swelling of the tentacle. Infected snails can be identified by these lesions. A remarkable change caused in the snail intermediate host of a fluke is shown by snails infected by the miracidia of the fluke *Leucochloridium* (*Distoma*) *macrostomum*. The sporocysts of this fluke develop in the tentacles of the snail and so affect them that they cannot be retracted by the snail. They are also coloured by red and green bands, so that they are conspicuous. It is possible that these coloured tentacles attract the birds which are definitive hosts of the fluke and which infect themselves by eating the infected tentacles. It has been observed also that the larvae of the filarial heartworm of the dog, *Dirofilaria immitis*, may, when they are numerous in the species of the mosquitoes belonging to the genus *Culex*, which are among their intermediate hosts, destroy the renal organs (*Malpighian tubules*) of these insects, so that 50 % of the insects may die. The microfilarial larvae of some other filarial roundworms, such as Bancroft's filarial worm, do less harm, probably because they live in the thoracic muscles of the many species of mosquitoes which are their intermediate hosts, and injury to these muscles disturbs the health of these hosts less seriously than injury to the renal organs would.

CHAPTER 9

OTHER IMPORTANT ASPECTS OF THE HOST-PARASITE RELATIONSHIP

In the preceding chapters we have discussed the routes by which parasitic animals enter and leave their hosts, some selected life histories, the effects of parasitic life upon the parasitic animal and some of the effects which the host may suffer. There remain for consideration some aspects of the host-parasite relationship which are either in themselves biologically interesting or are important because they show us how parasitic animals spread from one host to another and therefore help us to control the harm done by species which injure man and his civilisation. Some of the methods by which man seeks to control certain species of parasitic animals which injure him and his interests will be discussed in Chapter 10. In this chapter we will consider some consequences of the entry of parasitic animals into their hosts through its mouth and through its surface, of their exit through its surface and of their attacks upon the host's respiratory, renal and reproductive organs.

A. Some Consequences of Entry through the Host's Surface

The host's surface may be penetrated

(1) by the unaided efforts of an infective phase of the parasitic animal; or

(2) by the introduction of some phase of the parasitic animal through the host's surface by another animal which

is, or is not, an intermediate or definitive host of the parasitic animal introduced.

(1) Infective phases which penetrate the host's surface by their own unaided efforts must be sufficiently active and muscular to do this. Usually they are larvae and their penetration may be helped by small, piercing stylets or by protoplasmic papillae at the head ends of the larvae. The miracidium (fig. 41) of the liver fluke, *Fasciola hepatica*, has a papilla which helps it to penetrate into the snail inter-mediate hosts of this species and the larva of the pork trichina-worm has a stylet at its front end, which helps it to penetrate through the wall of the small intestine of its host and later into the muscles in which it encysts. The metacercariae of some species of flukes possess similar stylets. Some kinds of larvae which make their way through the skin of mammals enter the small pits (follicles) in the skin out of which the hairs grow and thus do not need to bore through the outer, more horny layers of the skin. Once these outer layers have been penetrated, the progress of the larvae may be helped by backwardly directed spines on the larva's surface, which prevent it from slipping back. The nymphs and adults of the burrowing mite, *Sarcoptes scabiei*, which causes human scabies, have spines on their backs which help them in this way, and a similar function may be performed by the spines on the larvae of the warble-flies (fig. 32) and sheep nasal-flies. Penetration is also, no doubt, aided, and possibly it is also sometimes initiated, by the sensitivity of the larvae to touch stimuli, which awaken muscular movements that impel the larva to bore into any kind of surface with which its head end comes into contact. The larvae of roundworms show this kind of reaction, and it may impel them, and probably also the

miracidial larvae of flukes as well, to bore into the outer surfaces of their normal hosts. It may also, however, impel them to bore into the outer surfaces of hosts in which they cannot grow to maturity. In these hosts they die, but they may produce in them the reactions described below.

The surface layers of the host usually react to some extent when the larvae of parasitic animals penetrate through them. When the host is well adapted to the parasitic animal the reaction may be slight; but if it is not, irritation and inflammation are caused, with their attendant itching and reddening. If the larva stays and develops in the skin, this reaction progresses all round the invader. Thus the larvae of the African tumbu fly, *Cordylobia anthropophaga*, which is related to the blowflies whose effects are described below, hatches from the eggs laid in sand or on human clothing and tears a hole in human skin with its mouth hooks. Through this it disappears into the skin in a few minutes to cause the development around it of a painful skin reaction resembling a boil, out of which the mature larva emerges after about 8 days to pupate on the ground. The larvae of the human warble-fly, *Dermatobia cyaniventris* (*D. hominis*), cause swellings under the skin of man, cattle and other animals similar to those caused by the warble-flies of cattle described in Chapter 3. The female of this species is interesting because she uses blood-sucking arthropods to carry her eggs to the host in which the larvae are parasitic. Among the blood-sucking species thus used are the large mosquito *Psorophora* (*Janthinosoma*) *lutzi*, species of *Culex*, the stable-fly *Stomoxys calcitrans* and sometimes a tick belonging to the genus *Amblyomma*. The female tropical warble-fly cements 200 eggs or so to one of these blood-sucking species which carry them to the host whose blood they suck. The eggs then hatch and the larvae bore through the skin of this host as the larvae of the cattle warble-flies do. In the boil-like swelling that they provoke the larvae mature and then they drop to the ground to pupate. Rarely the larvae of the cattle warble-flies get into human skin and cause similar, but less troublesome, swellings. They cannot mature in human skin.

The reactions of human skin to the larvae of hookworms and cercariae of flukes are interesting and important. Both these kinds of larvae may cause an itching inflammation of the skin (*dermatitis*) or an itching rash, with or without inflamed spots or papules. If the host scratches to relieve the itching, the skin surface may be abraded and infected with bacteria, so that the condition is made worse. This kind of reaction is usually more severe when the host is not fully adapted to the parasitic animal. It is the expression of the host's efforts to get rid of the invader. If the host is one in which the parasitic animal has been living for a long time, so that host and parasite have learnt to tolerate each other, the skin reaction of the host is less pronounced.

The skin-penetrating larvae of the hookworms illustrate this. Thus the larvae of *Ancylostoma braziliense*, which is normally parasitic in dogs and cats, cause, when they bore through the skin of these hosts, little reaction to them. The adults of this species can, however, live in man, who is not well adapted to them. Human skin therefore reacts more strongly to its skin-penetrating larvae, which cause in human skin an inflammatory rash accompanied by inflamed papules. It is called *creeping eruption*, because it creeps along the skin as the larvae move along in the skin. Another form of creeping eruption of human skin may be caused by the larvae of the roundworm *Gnathostoma spinigerum*, which is normally parasitic in the stomach of the cat, tiger, leopard and dog, where it may cause the formation of tumours (see Chapter 7). The creeping eruption caused by the larvae of *Ancylostoma braziliense* may be contrasted with the reaction of the human skin to the larvae of the human hookworm *A. duodenale*, which has been parasitic in man since the dawn of history, so that he is more fully adapted to it. This skin reaction is called *ground itch*, but it is less severe than creeping eruption.

Like the hookworm larvae, the cercarial larvae of flukes may cause skin reactions when they penetrate the skin of their hosts, and these also are more severe when the cercariae belong to species of flukes which are not usually parasitic in the host concerned. The cercariae of the human blood flukes still cause an urticarial rash when they penetrate human skin, although, like the hookworms, they have been parasitic in man since very ancient times. But there are also, in certain parts of the world, cercariae of species of flukes which cannot mature in man, but can penetrate his skin and cause itching rashes. Thus people bathing in the Great Lakes of Canada and the northern United States, and in New Zealand, India, France and parts of Germany and Wales, may be afflicted by a dermatitis called *swimmer's itch*, which is caused by the penetration of human skin by the cercariae of flukes which normally mature in aquatic birds (e.g. ducks) and perhaps other vertebrates (e.g. mice). These cercariae come from species of freshwater snails belonging to the genera *Physa* and *Limnaea*, which are the intermediate hosts of these species of flukes. A similar dermatitis of man called *swamp itch* may be caused by the cercariae of the blood fluke *Schistosoma spindale*, the adults of which are normally parasitic in sheep and cattle in India, Malaya and South Africa. *Collector's itch* is a name given to a dermatitis of man reported from the neighbourhood of the University of Michigan. It may be acquired by biologists seeking for animals in muddy or swampy areas inhabited by snails which are intermediate hosts of flukes whose adults cannot live in man, although their cercariae can penetrate human skin.

(2) When the infective phase is introduced by some other animal, this animal may be:

(*a*) An animal which is not a host of the parasitic species, but the non-parasitic mother of the parasitic infective larva. This mother is usually an insect which either lays her eggs on the surface of the host's body and larvae hatch from these to live parasitically in the host's surface tissues, or she herself penetrates the host's surface with an ovipositor in order to plant either her eggs or her larvae in the host's interior.

(*b*) The intermediate or definitive host of the parasitic species. Certain species of anopheline mosquitoes, for instance, are the definitive hosts of the malarial parasites which they introduce into man by penetration of his skin.

(*a*) The number of insects which lay either their eggs or larvae on the surface of other animals or inside their bodies is considerable and only a few types of them can be considered in this book. They can be divided into:

(i) Species which lay either their eggs or their larvae on the host's surface, the larvae remaining only in the surface layers, and

(ii) species which penetrate the host's surface to implant their eggs or larvae deep inside the host's body.

(i) Among species which lay their eggs or larvae on or in the host's surface layers are several two-winged insects (Diptera). Their maggots cause the various forms of a serious disease of man and domesticated animals called *myiasis*.

This disease may result when the mother insect lays her eggs on the host's surface or in wounds or sores on it. The insects are likely to do this when the wounds are infected with bacteria and are full of pus; or when the intact skin of animals, such as the sheep, is soiled by urine or faeces the odours of which attract the flies to it.

Among the insects whose maggots may infect the skin in this way are the blowflies, which include the bluebottle and greenbottle flies and the black blowfly, *Phormia regina*, which frequently lays its eggs in wounds infected by bacteria. But some other species, such as the flesh-flies whose larvae normally live on decaying flesh or similar material, may lay their eggs on wounds and sores. Among them are species of the genus *Wohlfahrtia*, which is a common pest of man and animals in south-east Russia.

A serious form of myiasis produced in this way is the disease of sheep known as 'blow' or 'strike'. 'Blow' is the name given to the laying of eggs by the flies. 'Strike' refers to the development of the maggots. The maggots, which are sometimes called 'wool-maggots', are those of species of blowflies belonging to the genera *Calliphora*, to which the bluebottle belongs, and *Lucilia*, which includes the green-bottle flies. The female flies are attracted to the sheep by the odours produced by bacterial action in urine or faeces adhering to the wool or in blood clots and scabs at the sites of bites inflicted by ticks. The breech of the sheep is often attacked, because the wool here is often soiled and clotted with excreta; but other parts of the body are also attacked, especially if the sheep sustain injuries of any kind that cause bleeding.

Closely related to the blowflies which cause strike are the 'screw-worm' flies which attack cattle. These belong to the genera *Cochliomyia* and *Chrysomyia*. *Cochliomyia americana* attacks cattle in North and South America. *Chrysomyia bezziana* attacks cattle in Asia and Africa, and also man. The larvae of this latter species are especially troublesome to man in parts of India. It also attacks animals there and, because its larvae need living tissues as food, it is a dangerous species, which cannot be controlled by the disposal of dead carcasses, as species which feed on dead flesh can. Its larvae may enter the nose and get into the frontal and ethmoidal sinuses and may even attack the bones in this region. The inflammation they cause may lead to inflammation of the membranes round the brain (meningitis) or to septicaemia. Its effects are thus similar to those caused by the sheep nasal-fly described below. The maggots of the species just described feed upon living flesh and are therefore truly parasitic. Those of another American species, *Cochliomyia macellaria*, and those of the Old World species *Chrysomyia megacephala*, feed upon dead carcasses, although the eggs of these species may be laid in sores upon living animals.

Myiasis may be caused by flies which lay, not eggs, but larvae upon the host's surface. An example of them is *Auchmeromyia luteolata*, which is related to the house-flies. The larva of this fly is called the Congo floor maggot. It hatches on the floors of huts or in sand and enters the skin of people who sleep on uncovered floors or on the bare earth.

Up to this point we have considered the penetration of the host's skin by the young only of the parasitic animal.

But the adult also may penetrate it. Thus the adult female chigger flea, *Tunga penetrans*, may burrow in human skin. As it grows and becomes more and more distended with eggs it may reach the size of a pea and cause much pain and inflammation. If its burrows become infected with bacteria, abscesses may develop and these persist after the flea has left the skin. Other species whose adults penetrate the skin of the host are the mites which cause the various forms of scabies described elsewhere in this book.

(ii) Examples of species which penetrate the host's surface to implant their eggs or larvae deep in the host's body will also be selected from the insects. The most striking instances are found among the Hymenoptera, to which order the ants, bees and wasps belong. All the members of this order possess a very efficient egg-laying instrument called the *ovipositor*, which is a modification of the hind end of the insect's abdomen. Ovipositors may have a piercing, sawing or stinging action, and they can penetrate such resistant structures as the bark of trees, so that it is not surprising that they can penetrate the surface layers of other animals.

Many species of Hymenoptera implant their eggs inside the bodies of other parasitic insects, so that they provide good examples of *hyperparasitism*. Some of them are useful to man because they plant their eggs in the bodies of insects which injure plants valuable to man. Thus most of the species of the superfamily Chalcidoidea lay their eggs on either the eggs, larvae or pupae of injurious butterflies, moths, bugs, two-winged flies and scale insects. Among them are members of the family Encyrtidae, which is composed entirely of species parasitic upon the eggs, larvae and pupae of other insects. An example of them is

Ageniaspis (fig. 113), which lays its eggs in the caterpillars of the ermine moth. When the larvae derived from these eggs have consumed the internal organs of the caterpillar, they emerge to form cocoons outside the caterpillar's remains. This caterpillar is attacked by at least five other species of parasitic Hymenoptera, including species of the genera *Apanteles, Pimpla* and *Angelia*, the last-named being itself attacked by an ichneumon-fly.

Some of the smaller ichneumon flies, such as species of the genus *Apanteles*, which belongs to the

Fig. 113. Cocoons of the encyrtid fly, *Ageniaspis fuscicollis*, in the caterpillar of the small ermine moth, *Hyponomeuta* sp., on whose tissues the larvae of *Ageniaspis* have fed

family Braconidae, lay many eggs in each host caterpillar, and their larvae bore their way out of their hosts to spin their own cocoons near by, so that the pupa of the host is found surrounded by many small cocoons of the insect that has been parasitic on it. Probably some readers will have found these yellow cocoons round a chrysalis in their gardens, and some of them, mistaking them for eggs, may have wondered how a pupa could have produced eggs. *Apanteles* itself does not escape being parasitised by hyperparasites which may destroy its larvae. *Microbracon mellitor*, a useful parasite of the cotton boll-worm, is mentioned in Chapter 10, but its eggs are laid on the surface of this caterpillar and the larvae of the parasite suck its body fluids.

Some large ichneumon flies, such as species of the genera *Ophion* and *Pimpla*, lay their eggs inside caterpillars, one egg in each. The larva of the ichneumon may leave the caterpillar host before its cocoon is spun, or only when the cocoon is ready. The host caterpillar then dies and the parasite spins its own cocoon inside that of the host, so that the adult parasitic insect must emerge through the walls of two cocoons.

Two other families of Chalcidoidea whose females lay their eggs, not upon the larvae, but upon the eggs of other insects by means of ovipositors are biologically interesting. These are the Trichogrammidae and the Mymaridae (fairy-flies), all the species of which are very small.

One species of fairy-fly, *Alaptus magnanimus*, is probably the smallest insect known; it is only about 0·2 mm. long. *Polynema natans*, another fairy-fly, has the remarkable habit of swimming with its wings beneath the fresh water in which its hosts, which include water-boatmen (*Notonecta*) and some dragon-flies, live. This fairy-fly lays its eggs upon the eggs of the common dragon-fly, *Calopteryx virgo*, which are laid upon the leaves of white water-lilies belonging to the genus *Nymphaea*. One egg only is laid in each dragon-fly egg. In a few days the whole egg is consumed and the parasitic larva pupates. *Prestwichia aquatica*, on the other hand, which is one of the Trichogrammidae, swims with its legs and is parasitic upon the eggs of water-boatmen, water-scorpions, pond-skaters and water-beetles. Another hymenopterous insect which enters the water to lay its eggs upon the host which nourishes its larvae is *Agriotypus*, which lays its eggs upon the larvae of caddis-worms.

Some of the Trichogrammidae, however, attack the eggs of insects which are not aquatic. Thus *Trichogramma evanescens* lays its eggs in the eggs of insects belonging to the orders Lepidoptera, Diptera, Hymenoptera, Hemiptera and Neuroptera. This minute insect examines and selects hosts upon which its offspring can develop, and the defences of its hosts against it include the physical and chemical composition of the shells and contents of their eggs. This remarkable small insect also seems to know when one of the host's eggs has been parasitised by a female of its own species, for it does not lay in the host eggs thus, so to speak, already occupied.

Just as remarkable is the fact that the eggs of some insects, such as those of the sawfly, *Selandria sixi*, stimulate the ovipositor of *Trichogramma evanescens* in such a way that the impulse to lay eggs is inhibited and the ovipositor is withdrawn before a *Trichogramma* egg is laid.

Turning from the Hymenoptera to the two-winged Diptera we find that the larvae of all the Tachinidae are parasitic. Usually their hosts are the larvae of Lepidoptera and, less often, those of Hymenoptera; they may also parasitise the larvae and adults of Orthoptera, Coleoptera and Hemiptera. They deposit their eggs either on the host or on its food plants. Some species of tachinids lay, not eggs, but living larvae produced inside the mother insect, and these may be laid either on or under the integument of the host. Thus *Ernestia rudis* lays them on the skin of *Taeniocampa stabilis*. Inside their hosts the larvae of Tachinidae feed upon various organs, many of them feeding upon different organs at different times during their life histories and showing adaptations to the phases of the host's life history that are correlated with these feeding habits. At some time during their life histories they must breathe air, either through holes made in the host by themselves or by their mother insects that introduced them, or through connections established between the external world and their air tubes (*tracheae*). The host reacts to their effects by forming sheaths derived from its skin or air tubes, which surround the tachinid larvae.

(*b*) The introduction through the skin of the infective phase of a parasitic animal by a host which is either the intermediate or definitive host of the parasitic animal introduced has been described in Chapter 8. When it occurs, the actual entry of the parasitic animal depends only partly upon its own efforts. The most it can do to help its entry is to place itself inside its intermediate or definitive host in a position favourable to its exit from this host. The sporozoites of the malarial parasites, for instance, can do no more for themselves than to find, and assemble in, the proboscis of the mosquito. The infective larvae of filarial roundworms and those of species transmitted in a similar way, such as *Habronema*, do a little more for themselves, because they actually break out of the proboscis when their insect transmitters suck blood and then bore through the skin of the host in which they become parasitic. But the entry of none of these infective phases could be effected without the aid of the blood-sucking transmitter.

The habits of the blood-sucker may therefore have an important, or even a decisive, influence on the survival of the parasitic animal. Their influence on the distribution of the malarial parasites is indicated in Chapter 10. They are certainly decisive when the blood-sucking vector transmits the parasitic animal mechanically in the manner described in Chapter 8. In that chapter it is explained that the irritation caused by the bite of the blood-sucker may also help the transmission of the parasitic animal because it causes the host to scratch and rub into the bite the parasitic animal present in the excreta of the vector.

The habits of the vector may also influence the sites in the body of the host occupied by the parasitic animal that it introduces. In Central Africa, for instance, the blackfly,

Simulium damnosum, which is the intermediate host of the filarial roundworm, *Onchocerca volvulus*, bites human beings chiefly low down, on the legs or lower parts of the body, so that a naked baby placed upon a table may escape its bites. For this reason the nodules caused by the host's reaction to the infective larvae of this roundworm are most often found on the lower parts of the bodies of African natives. In Guatemala and Mexico, on the other hand, other species of *Simulium* which are the intermediate hosts of the same species of roundworm bite human beings chiefly on the head and upper parts of the body, so that the nodules caused in this part of the world are less often on the lower half of the body.

B. Some Consequences of Exit through the Host's Surface

Parasitic animals do not usually cause disease by their method of leaving the body of the host, but two serious diseases of man are caused in this way. The spined eggs of the human blood flukes, *Schistosoma haematobium*, cause, as they make their way through the walls of the bladder, chronic inflammation, hyperplasia, fibrosis and the formation of papillomata. The eggs are surrounded by the type of tissue reaction called a worm nodule described in Chapter 7. All these changes interfere with the function of the bladder. The spined eggs of *S. mansoni* cause similar reactions in the walls of the large bowel, the results being dysentery, diarrhoea and other complications.

The guinea-worm, *Dracunculus medinensis*, provides another example of the consequences of exit from the host through its surface. This roundworm may cause no apparent symptoms during the 10–14 months during which it is maturing after its entry through the mouth of the host in the *Cyclops*, which is its intermediate host. But when the adult female is ready to liberate to the exterior the living larvae in her uterus, she approaches the skin of the host, appearing usually under the skin of the sole of the foot or that of the ankle, although she may 'point' on the trunk, buttocks, arms or breast, or may cause blindness by appearing in the orbit. When she thus 'points' under the skin, or a few hours before this, there may be local reddening of the skin, itching, generalised rash, slight fever, nausea, vomiting, diarrhoea, asthma-like symptoms or even giddiness or syncope. These symptoms are probably caused by the liberation by the worm of poisonous substances into the host. When the head of the female worm comes to the surface, or a few hours before it does, itching and boring, dragging or burning pain are felt at this point. A small blister appears, which may grow till it has a diameter of 2–6 cm. If this blister is immersed in water, it bursts and the head of the worm protrudes through a small hole in the skin. Douching the area with cold water at this time will cause the extrusion of the uterus of the worm by rupture of the worm's head or through its mouth, and a milky fluid is exuded from the uterus, which contains the larvae. Repeated douching will 'milk' the uterus until all the larvae have been discharged.

The difficulty of removing the female worm, which is firmly enclosed in the reacting tissues around it, and the method of doing it by rolling it on to a stick, are well known. The wound caused may become infected with bacteria, so that inflammation and great pain may result and the tissue reactions that then follow may cripple the host. The fact that immersion of the affected part in water causes the liberation of the larvae, which must enter water in order to find the crustacean *Cyclops*, which is the intermediate host, is biologically very interesting.

C. Some Consequences of Entry through the Host's Mouth

Parasitic animals which use this portal of entry may lie in wait in the substance of the solid food of the host until this is eaten by the host. This is the method used by the pork trichina-worm, whose larvae wait in the muscles of the host in which they have been produced until these infected muscles are eaten uncooked, or insufficiently cooked, by another host. It is also the method by which some definitive hosts infect themselves when they eat intermediate hosts infected with the larval stages of tapeworms and flukes. Man, for instance, may infect himself with beef and pork tapeworms when he eats beef or pork infected with the bladderworms of these species. The second intermediate host also commonly infects itself in this manner by eating the first intermediate host.

Many parasitic animals, however, which enter their hosts through the mouth, are not already living in the substance of the host's food, but are merely mixed with it or are on its surface; or they are in the host's drink; or they are communicated to the host's mouth by some agency independent of themselves. The food and drink are, as we say, contaminated by the infective phases and, for this reason, this method of infection of the host is often called the *contaminative method*. The infective phases which use it are either passive eggs or cysts or other structures covered by protective envelopes, or they are motile infective larvae. They are all liable to be killed by unfavourable surroundings, but some of them can survive injurious external conditions for a remarkably long time, especially if they are kept sufficiently moist. They are often conveyed to their host in water that it drinks. Another source of infection with them is vegetable food growing in damp or wet places. Man may infect himself with them by eating aquatic plants, such as watercress, which are not cooked before they are eaten. Human beings, for example, on the rare occasions when they infect themselves with the liver fluke of sheep and cattle (*Fasciola hepatica*), usually do so by eating uncooked watercress or some similar salad plant on which the cercariae of this species of fluke have become encysted.

The water used for the irrigation of crops and gardens may also be a source of infection, especially in dry countries in which elaborate irrigation systems are constructed. A good example of the infection of human beings through water used for the irrigation of gardens and allotments has been revealed by Russian parasitologists, who investigated the causes of the infection of people in Moscow with the roundworm *Ascaris lumbricoides*. These workers found that the eggs of this roundworm were not all killed by the passage of human excreta through the sewage works, but that many of them survived and passed on into the water of the river Moskva, into which the sewage effluent was discharged. The water of the river Moskva is used by gardeners in Moscow for watering the plants in their allotments, and living eggs of *Ascaris* capable of infecting man were found on the cabbages and other plants growing in the allotments. A complete cycle of infection thus carried the eggs from the houses through the river to the gardens and so back to the people from which they came. Similar cycles have been studied in Odessa and in the Darmstadt area of Germany.

In Stuttgart, on the other hand, it has been shown that the modern sewage system used there destroys practically all the eggs of *Ascaris*. In countries in which human excreta are used to manure plants that are eaten uncooked, the eggs of animals parasitic in man may be readily transmitted from man to man by this route. *Ascaris lumbricoides* frequently spreads in this way from one human being to another.

Many intermediate hosts are infected with the larvae of parasitic animals by contamination of their food or drink with infective phases derived from the definitive host. Thus dogs, which are the definitive hosts of the tapeworm *Echinococcus granulosus*, may deposit excreta which contain the eggs of this species upon lettuces and other garden plants that are eaten uncooked by man, so that man may eat the eggs and so become the intermediate host in which the bladderworms (hydatid cysts) of this tapeworm develop.

Another way in which food and drink may become infected is by contamination of it by infective phases of parasitic animals which either (*a*) adhere to the feet of insects, birds, rats and other animals or (*b*) exist in the excreta of these animals which are deposited by them upon human food or drink. Eggs thus deposited in human food or drink may be either those of species parasitic in the animals which deposit them or mere passengers through the food canals of these animals, in which they are not injured or killed. It is known, for example, that eggs of *Ascaris lumbricoides* and those of the whipworm *Trichuris trichiura*, both of which infect man, can pass unchanged through the food canals of cockroaches, which may then deposit them upon human food. The eggs of the human hookworms can also be carried about by certain species of wasps, which transport them from place to place, not in their food canals, but on their wings, bodies, legs and mouth parts, to which the eggs adhere. Other animals, including some household pests, may also contaminate human food in this way. The common house-fly is a well-known culprit in this respect. Even foods and drinks which have been cooked or otherwise freed from infective phases of parasitic animals may be again contaminated after they have been disinfected and before they are eaten or drunk. The utensils, also, with which human beings eat and drink may be contaminated in similar ways. There are, indeed, so many ways in which infective phases of parasitic animals may be conveyed to the food and drink, or otherwise to the mouth of the host, that a full account of them would be tedious. The reader will be able to think them out for himself if he is guided by the life histories described in Chapters 3 and 4.

Apart from the entry into the host's mouth with the food or drink, there are at least two other methods by which the infective phases of parasitic animals may be transmitted to the host's mouth. They may be passed to the mouth by the fingers or other parts of the hands, or they may pass in with air that is inhaled.

Three important parasitic animals which may be transmitted to the human mouth by the hands or fingers are the pork tapeworm, the hydatid tapeworm and the human seat-worm (pinworm or threadworm). It is the eggs of the tapeworms which are transmitted to the mouth. The eggs of the pork tapeworm may come either from a tapeworm of this species present in the food canal of the person who gets its eggs on to his fingers and swallows them or from the excreta of another person. The eggs of the hydatid tapeworm cannot come from man, because this tapeworm cannot live in human beings; they can come only from the dog or one of this tapeworm's other hosts. Its eggs may get on to the hair of the dog and thence on to the fingers and hands of people who fondle or stroke the dog. The dog may also lick the eggs off its hair and then transfer them to human beings whom it licks, or even to plates and other utensils used for human feeding, which, if they are not well washed, may transmit the eggs to the human mouth.

The sticky eggs of the human seat-worm, *Enterobius vermicularis*, are well adapted to transmission on the hands and fingers. The worms lay their eggs, chiefly at night, around the anus and cause irritation as they do so, with the result that this area is scratched and the sticky eggs get on to the fingers or under the finger nails and by these are transferred to the mouth. When hands or fingers thus contaminated with them handle door-knobs, stair-rails and other pieces of furniture, or various household objects, the eggs adhere to these and may be picked up by the hands or fingers of people uninfected with the worm. Infection with the seat-worm can be acquired in these ways only if the eggs are picked up soon after they have been laid and if they have not been killed by drying. They are not very resistant, and sunlight and ventilation, which encourage drying of them, kill them fairly quickly. Their entry into the mouth with air that is inhaled is considered below.

The entry of the infective phases of parasitic animals with air that is inhaled is much less common than we might expect, largely because the infective phases cannot be transported in the air unless they are dried to such an extent that the drying kills them. Nor do animals which are parasitic in the air-breathing organs often enter these organs by this method, although bacteria and other kinds of parasites frequently do. Some of the infective phases of some species parasitic in the food canals of their hosts may, nevertheless, be able to withstand drying sufficiently to be blown about in the air and to be taken into the mouth with the air and then swallowed.

It is known that the infective larvae of some species of roundworms parasitic in the food canals of domesticated animals can survive drying to this degree and they have been found alive in the dust of farm buildings, with which they may be inhaled and swallowed. A proportion of human beings who become infected with the eggs of the human threadworm also infect themselves by taking in air in which these eggs are suspended. The number of persons who thus infect themselves is, however, relatively small, because, although many eggs of this species may be found in the dust of schoolrooms and other places inhabited by large numbers of children, the eggs of this species cannot resist drying very well, and most of the eggs found in the dust and air are dead. The air beneath the bedclothes is, however, relatively warm and moist, and in it, as Dutch biologists have found, human threadworm eggs may remain alive and may be inhaled during the night in relatively large numbers. It seems likely, therefore, that the air and dust of schoolrooms or the home are less likely to be sources of infection with the eggs of the human threadworm than the space under the bedclothes, in which a child may spend half its life. In this warm situation, moreover, the eggs may hatch on the skin round the anus and it has been shown that, when this happens, the larvae can re-enter the human bowel through the anus and develop normally in it. In this manner a repetition of the infection (relapse) may be set up. This probably explains why light infections with this worm sometimes persist, with intervals of apparent freedom from them for a few months.

D. Some Species which Attack the Respiratory Organs

When we remember how readily bacteria and viruses enter the external openings of the respiratory organs of animals and cause diseases of them, it is perhaps natural to expect that parasitic animals which attack these organs also frequently enter them by way of their external openings. It is, however, an interesting fact that relatively few animals whose adult stages are parasitic in the respiratory organs of animals enter their hosts by this route. Among those that do is the mite, *Acarapis woodi*, the cause of acarine disease of bees, which enters through the spiracles of the bee that give access to the bee's breathing tubes (*tracheae*). The mites live in these small tubes, piercing their walls to suck the tissue fluids of the bee, and when they are numerous they block up the breathing tubes, depriving the bee of oxygen, so that it may lose its power of flight. A harmless mite which lives in the air sacs and respiratory passages of poultry, but usually does little harm to them, is *Cytoleichus nudus*.

But, on the whole, parasitic animals, when they injure the respiratory organs, do not enter these organs by way of their external openings. Certainly the species which do most harm to the lungs of mammals enter by way of the host's mouth and reach the lungs after penetration of the bowel wall. Usually their eggs are produced in the host's lungs or air passages, and either they or the larvae which hatch out of them are coughed up and swallowed into the host's food canal. They are then passed out of the host in its droppings, and are either then eaten by another host, in which event the life history is direct, or they enter an intermediate host, which is eaten by a definitive host. Passing thus either directly or indirectly into the food canal of the new mammalian host, the larvae bore through the wall of this canal and find their way to the respiratory organs, in which they grow to maturity.

The 'gapes' worm *Syngamus trachea* (fig. 111), of poultry, turkeys, geese and various wild birds, which is parasitic in the windpipe and causes cough, shortness of breath and spasms of asphyxia which may result in the death of the host, has a life history of this kind. Another species of this genus, *S. laryngeus*, may be parasitic in the voice-box (*larynx*) of cattle in India, Malaya and South America. It causes similar effects and occasionally it gets into the larynx of man. *S. nasicola* may be found in the nasal cavities of cattle, sheep and goats in Brazil, Africa, Turkestan and the West Indies.

A species which is parasitic in the nasal cavities of the dog, fox, wolf and occasionally in those of the horse, sheep, goat and man is the arthropod species mentioned in Chapter 5, *Linguatula serrata*, which is often called the tongue-worm, because its shape recalls that of the mammalian tongue. It is not a worm, but is related to the scorpions, spiders and ticks (Arachnida). The hosts just mentioned are its definitive hosts. Its eggs pass down the food canal of the definitive host and are passed out with its droppings. They are eaten by the intermediate hosts, which may be sheep, cattle, horses, rabbits and possibly also other mammals. In the food canal of the intermediate host the larvae emerge from the eggs and bore into the lymphatic glands near to the food canal, in which they cause tissue reactions, some of which may resemble those of tuberculosis so closely that it is difficult to distinguish between the two by naked-eye examination. Later the larvae may migrate to the liver, lungs and other organs of the intermediate host. They remain in the intermediate host until it, or its infected organs, are eaten by the definitive host, into whose nasal cavities they then make their way. Relatives of the tongue-worm are parasitic in the lungs of snakes and some other reptiles in the tropics, and some of these may use man as an intermediate host. When they do, their larvae may cause serious damage to the lymphatic glands or other organs of man.

The nasal cavities and the sinuses at the back of the nose which communicate with them are the habitat of the larvae of the sheep nasal-fly, *Oestrus ovis*, which is a relative of the warble-flies of cattle and of the horse bot-flies. These larvae are produced inside the body of the female fly and are laid directly into the nose or near it. After a period of life inside the nose, near its external openings, the larvae crawl farther up the nasal passages and may get into the sinuses, where they cause inflammation (*sinusitis*). They feed upon the tissues and may even attack the bones and cause inflammation of the membranes round the brain (*meningitis*). They have been known to get into the brain itself and into the eye. They develop during the winter following their introduction into the nasal passages and leave the host during the following spring to pupate upon the ground. In some parts of the world they cause serious loss to the sheep farmer. The harm that they do resembles that done by the larvae of *Chrysomyia bezziana* mentioned above.

Rhinoestrus purpureus, the Russian gadfly, is another species belonging to the same family as the sheep nasal-fly.

Parasitic lower down the air passages in the air tubes leading to the lungs are the roundworms which are often called the lungworms, some species of which cause serious diseases of sheep, cattle, horses, pigs and other animals. Some of them pass directly from one host to another, others require an intermediate host. Direct transmission is characteristic of all the species of the genus *Dictyocaulus*.

One species of this genus, the large lungworm of sheep, lives in the bronchi of sheep, goats and some wild animals which chew the cud (*ruminants*); another species lives in the bronchi of cattle and deer and another in those of the horse and its relatives. The eggs of these species are usually coughed up from the bronchi and are then swallowed, although some of them may hatch in the air passages. When the eggs are swallowed they hatch in the food canal, and the larvae which come out of them are passed out of the host with its excreta. They survive for a while outside the host, and if they are eaten by another suitable host, they penetrate the walls of its intestine and migrate to the lymph glands near to the food canal. In these they moult their skins and then pass, by way of the lymph and blood streams, into which they penetrate, to the small capillaries of the lungs, out of which they make their way to become parasitic in the bronchi.

These species do not therefore avail themselves of what would seem at first sight to be an easy way out of the host, with its expired air or with mucus sneezed or coughed out, and an easy way into the host, with its inspired air. They use, instead, the food canal as a way out and in and depend upon the migration of their larvae from the food canal to the lungs by a route which is similar to that used by the migratory larvae of such species as the large roundworm of the pig. A species with a life history similar to that of *Dictyocaulus filaria* is *Capillaria aerophila*, a relative of the whipworms, which is parasitic in the trachea and bronchi and rarely in the nasal passages of foxes, dogs, cats and deer relatives.

Among the lungworms whose transmission from host to host is indirect after a period of development inside an intermediate host there are species which cause serious diseases of farm animals. Thus the species *Protostrongylus rufescens* lives in the smaller air tubes (*bronchioles*) of sheep, goats, deer and possibly also rabbits and hares. Its intermediate host is a species of land snail belonging to the genus *Helicella*. Another species, *Muellerius capillaris*, lives, not in the air passages, but in the tissues of the lung. Its intermediate hosts belong to the genera of land snails, *Helix* and *Succinea*, and to three genera of slugs, *Arion*, *Limax* and *Agriolimax*. The lungworm of pigs, *Metastrongylus elongatus*, lives in the pig's bronchi, and its intermediate hosts are earthworms belonging to the genera *Diplocardia*, *Eisenia*, *Dendrobaena*, *Lumbricus* and *Helodrilus*.

The larvae of these species of lungworms which use intermediate hosts, like those of species which pass directly from one host to another, pass out of the definitive host in its droppings and enter the intermediate host. The unknown life histories of the lungworm of the cat, *Oslerus osleri*, and the dog, *Crenosoma vulpis*, are probably similar. A roundworm species which attacks neither the air tubes nor the tissues of the lungs, but the blood vessels of these organs, is *Aelurostrongylus abstrusus*, which lives in the smaller branches of the artery that takes venous blood to the lungs of the cat (*pulmonary artery*). Its larvae hatch out of the eggs in the capillaries of this artery, make their way into the air sacs to which this artery is carrying venous blood to be aerated and thus reach the air passages, up which they travel to the junction of the trachea and the gullet. They then pass over into the gullet and are swallowed into the food canal. They are passed out of the cat in its droppings and must then enter a mouse. In the connective tissue under the skin and between the muscles of the mouse these larvae develop inside small cysts. After a development in these cysts which lasts for about 3 weeks, the larvae are able to infect cats which eat the mice infected with them. Another species, *Angiostrongylus vasorum*, lives in the pulmonary artery and right ventricle of the heart of dogs and foxes. Its intermediate host is not known.

Among the flukes parasitic in man, a species which offers a parallel to the life history of the roundworms just described which use intermediate hosts is *Paragonimus westermanii*. This species lives in cysts in the lungs and bronchi and occasionally in the liver, spleen and some other organs of man, the dog, the wolf, the tiger, the rat and the pig. Its eggs are laid in these cysts. They pass into the air passages either by channels which connect the cysts with these passages or by rupture of the cysts into them. The eggs are then either coughed out or are swallowed and passed out with the excreta. When they have left the definitive host in either of these ways, they can develop further only in various species of snails. Most often they use species belonging to the genus *Melania*. In these intermediate hosts cercariae develop which escape into water, and these cercariae must then penetrate into a crab or a crayfish, of which several species are suitable. The definitive host infects itself by eating the infected crustacean after the cercariae have developed in them into infective meta-cercariae. This species, therefore, like the fish tapeworm of man, *Diphyllobothrium latum*, requires two successive inter-mediate hosts, but its second intermediate host, unlike that of the fish tapeworm, does not eat the first one.

E. SOME SPECIES WHICH ATTACK THE RENAL ORGANS

It is probably true to say that the majority of the parasitic animals which attack the renal organs (kidneys) of mam-mals do not enter their hosts by way of the external openings of these organs, but by some other route. We do not yet know the route of entry of many species which are para-sitic in the renal organs of other kinds of animals, whether these be the kidneys of the lower vertebrates, the mal-pighian tubules arranged in a ring around the hinder end of the food canals of insects and their relatives, or the various kinds of kidney tubules of other invertebrate animals; but we do know that some of them enter by the external outlets of these organs and some by other routes.

One interesting species which does not enter by the outlet of the renal organs is the fluke *Polystoma integerrimum*, (fig. 75), for instance, whose life history is described in Chapter 6. It enters its host, the frog, by way of the gills of the tadpole stage and only later finds its way to the bladder of the adult frog. The human blood fluke, *Schisto-soma haematobium*, the spined eggs of which cause severe chronic inflammation and other tissue reactions in the walls of the bladder of man, enters its host by way of the mouth or through the skin. The many species of coccidian Protozoa which are parasitic in the renal organs of shell-fish and other invertebrates and in those of rats, mice and other rodents and in those of birds and other vertebrates enter their hosts through the mouth; and there are species of roundworms which enter their hosts either through the mouth or skin, or by both these routes, and then severely damage or even completely destroy the kidneys of mammals. Because these species of roundworms may infect domesticated animals or man, they are economically important. Thus *Capillaria* (*Trichosoma*) *plica*, a relative of the whipworms, attacks the urinary bladder and sometimes the kidney of foxes, dogs and wolves. Its eggs pass out in the urine, and the infective larvae develop in the open and are swallowed by the new host. The eggs of *Stephanurus dentatus*, which is parasitic in the substance of the kidneys of pigs and in the fat around them, pass out of the host in its urine and the infective larvae, like those of the hookworms, enter other pigs either through the mouth or by penetrating through the skin. The larvae reach the liver by way of the blood and burrow through this organ to reach its surface, from which they find their way to the tube (*ureter*) which leads the urine away from the kidney and their activities cause the formation of cysts in the kidney or in the tissues around it. These cysts communicate with the ureter by fine canals and the adult roundworm lives in them.

Another roundworm which causes severe disease of the kidney or its complete destruction is the largest roundworm known, *Dioctophyme renale*, a giant species, the female of which may reach a length of 10 cm. It is parasitic in the kidneys of dogs, foxes, weasels and other wild carnivores, and occasionally also in those of horses, cattle, pigs and man. Its life history is not yet fully known, but probably it requires

an intermediate host. There is some evidence that this intermediate host can be a freshwater fish, and that one of the
intermediate hosts is the European freshwater fish *Idus idus*.
If this is true, this roundworm enters its definitive host
through its mouth when the intermediate host is eaten.
Among the insects an interesting species is *Melinda cognata*,
related to the blowflies. Its eggs are found in the mantle
cavity of the land snail, *Helix virgata*, and its first-stage
larvae develop in the kidney of this snail.

The route of entry into the hosts of a number of other
species which are parasitic either in the renal organs or in
their outlets to the exterior is not yet known, but certain
biological features of some of these are interesting. There
is, for instance, the roundworm species *Trichosomoides
crassicauda*, mentioned in Chapter 6, which lives in the
urinary bladder of the rat and whose diminutive males
are hyperparasitic in the wombs of the females of the same
species, three or four males being often found in the uterus
of a single female. There are the rare instances of the entry
of the maggots of two-winged insects into the urethra and
bladder of man. Usually these have been the maggots of
the lesser house-fly, *Fannia canicularis*, or those of its relative
the latrine-fly, *Fannia scalaris*. These authentic rare instances must, of course, be carefully distinguished from
records of maggots which have accidentally got into
receptacles used for human urine. There are also the
developmental phases of certain roundworms, such as the
larvae of the heartworm of the dog, *Dirofilaria immitis*, which
develop in the malpighian tubules of the mosquitoes
belonging to the genera *Aedes, Anopheles, Culex* and *Myzo-
rhynchus*, which are the intermediate hosts of this species.
Another species which is parasitic in the Malpighian

tubules of insects is the minute amoeba *Malpighiella refringens*, which lives in the cells lining the Malpighian tubules of the rat-flea, *Nosopsyllus fasciatus*, and the dog-flea, *Ctenocephalides canis*. Another minute amoeba has been found in the Malpighian tubules of the honey bee.

F. Some Species which attack the Reproductive Organs

The reproductive organs of animals are attacked by many species of parasitic animals. One reason for this is, no doubt, the fact that these organs provide all the nutritional elements which are necessary for the development of the fertilised egg of the host and are therefore rich sources of food for the parasitic animal.

Parasitic animals which attack the reproductive organs may enter by way of the ducts which discharge their products to the exterior or by reaching these organs from some other part of the host's body. They may live either in the ducts or in the reproductive organs themselves, or they may attack the larva or developing embryo of the host. Some of them are described elsewhere in this book and these will therefore only be mentioned in this chapter, in which we have space only for notes on a few representative and interesting species parasitic in the reproductive organs.

I. Species which enter through the Genital Ducts

These may be divided into

(1) those which enter these ducts and live their whole lives in them;

(2) those which enter these ducts and proceed from them to live in some other part of the host's body.

(1) *Species which Live their Whole Lives in the Genital Ducts*

Most of these are introduced passively into their hosts during the act of sexual congress. An example of them that causes abortion in cattle is the protozoon called *Trichomonas foetus*. This is a pear-shaped species, 5–15 microns long, which moves by means of three flagella free from its body and one attached to it by an undulating membrane similar to that of the trypanosomes. It is communicated to cows by bulls in whose sexual ducts it may be parasitic. It has also been transmitted experimentally to sheep. Trichomonads have been found in the genital tracts of horses and roe-deer.

Another species of this genus is *T. vaginalis*, which may be found in the vagina of women. It causes the discharge of a creamy, frothy, acid fluid. It is transmitted by men who are also infected. Relatives of these species live in the vaginal canals of monkeys, but these cannot live in man. Yet other relatives of them live in the food canals of man and other animals, and it has been suggested that the species which now live in the vagina and womb have been derived from species which lived in the food canal and entered the vagina by way of the anus, which is near the vaginal opening.

Another species of parasitic flagellate protozoon which lives in the genital ducts and is transmitted by coitus is *Cryptobia* (*Trypanoplasma*) *helicis*, which lives in the *receptaculum seminis* of species of land snail belonging to the genus *Helix*. Many species of this genus are common in Britain. Just before copulation occurs, the flagellates bore into the spermatophore which contains the sperms of the snail and are transferred with these to another individual. *C. isidorae*

is a closely related species which is parasitic in the receptaculum seminis of *Isidora tropica*, a South African air-breathing
snail which lives in fresh water; and *Cryptobia carinariae*
lives in the receptaculum seminis of the marine mollusc,
Carinaria mediterranea. *Cryptobia vaginalis* is parasitic in the
medicinal leech, *Hirudo medicinalis*. It lives, however, not in
the male reproductive organs, but in the vagina of this host.
Cryptobia dendrocoeli, on the other hand, is parasitic in the
epithelium and contents of the intestine of the planarian
flatworm, *Dendrocoelum lacteum*; but it also enters the reproductive organs and various glands and connective tissue
cells and may, according to some authorities, be transmitted
by copulation. The species of *Cryptobia* transmitted by
leeches are mentioned in Chapters 5 and 8.

Another species which is, according to some observers,
transmitted by copulation is the nematode *Rhabdias* (*Angio-
stoma*) *helicis* which lives in the oviducts of land snails
belonging to the genus *Helix* and also in their seminal
vesicles; it may also enter the hermaphrodite genital gland
itself.

The small males of the nematode species, *Trichosomoides
crassicauda*, which live in the vagina of the females of the
same species have already been mentioned. Aberrant individuals of some species of parasitic animals may wander
into the walls of the genital ducts and cause the formation
of abscesses in them. Thus, abscesses of the human uterus
and vagina may be caused by aberrant individuals of the
nematode species, *Ascaris lumbricoides*, which may also enter
the urinary bladder and urethra of the male.

(2) Species which Enter the Genital Ducts and then Pass to other Parts of the Body

We must be content with one or two examples of these. An important one is *Trypanosoma equiperdum*, which causes the serious disease of horses, asses and mules which is called dourine. This disease caused much trouble in the East during the war of 1939–45. This species of trypanosome is normally transmitted from host to host by coitus. After it has thus been introduced into the genital tract of the female, it causes oedema and swelling of the mucous membrane and other tissue reactions in the walls of the genital tract. It may make its way into the blood, but is usually not present there in large numbers. *T. equiperdum* is thus a species of trypanosome which has become independent of transmission by the blood-sucking intermediate host which trypanosomes normally employ. It has been suggested that it is a strain of *T. evansi*, which is the cause of the disease called surra in horses and their near relatives, camels, and some other mammals, and is transmitted to them by tabanid blood-sucking flies.

Another species of trypanosome which can be transmitted from one vertebrate host to another by coitus, but passes beyond the genital ducts, is *T. cruzi*, the cause of the South American trypanosomiasis which is called Chagas's disease. This species, which is normally transmitted through the skin in the manner described in Chapter 8, and can also be transmitted by the human mother to her baby in the mother's milk (see Chapter 7), has been transmitted experimentally to female mice by the introduction of it into the vagina.

II. Species which are Parasitic in the Reproductive Organs (Testes and Ovaries)

There are so many of these that only a few of them can be mentioned. Some of them reach the reproductive glands by entry through the genital ducts, but most of them enter the host by other routes and then proceed to infect its reproductive glands. They can be divided into those which injure the sex cells (*gametes*) and those which do not.

(1) *Species which do not Injure the Gametes*

A well-known example of these is the gregarine protozoon *Monocystis agilis* which lives in the seminal vesicles of the earthworm *Lumbricus terrestris*, where it feeds on the protoplasm to which the developing sperms are attached. We do not know how this species passes from one earthworm to another. It is not found in the cocoon in which the fertilised eggs of the earthworm are enclosed, so that it is unlikely that it is transmitted during copulation. Because its infective sporozoites are enclosed in a resistant envelope called the *pseudonavicella*, it seems likely that the life history includes a sojourn in the external world and that the earthworm infects itself by eating these pseudonavicellae, the sporozoites being liberated by the action of the digestive juices of the earthworm to find their way into the mass of protoplasm from which the sperms develop.

(2) *Species which do Injure the Gametes*

An adequate description of these would occupy the greater part of this book. We must confine ourselves to brief notes about a few species whose parasitic life illustrates important principles. One interesting feature to be noticed is the fact that some of these species do not destroy or injure the egg, but use it as a means by which they are transferred to the larva and adults of the succeeding generation which develops from the infected egg.

Some of the species which injure the sex cells of both sexes of the host are considered in Chapter 8, in which the remarkable effects of their destruction of the organs which produce the sex cells are considered. A species which attacks the genital organs of mammals is *Cuterebra emasculator*, one of the bot-flies, which injures the scrotum of male squirrels, rabbits and field-mice. Some species which live upon the host's eggs, either while these are still inside the mother or after they have been laid, or upon the larvae which hatch out of the eggs, are considered earlier in this chapter. Here let us consider the important and biologically very interesting fact that some species of animals parasitic in the female adult host enter this host's ovaries and either damage the eggs in these or remain in them to infect the generation of the host developing from them, or use them as a means of transport from one host to another. This process is nowadays known as *transovarian infection* of a host. This term is a useful one, because it distinguishes clearly the class of parasitic animals which enters the egg and then infects the offspring which develop from the egg and differentiates them from parasitic animals which pass, not through the egg, but from the tissues of the mother to those of the young developing inside her body. This latter method of infection is called *transplacental infection*. Both transovarian and transplacental infection are sometimes called hereditary infection; but this term, if it is used at all, should refer only to the transmission of disease, or the liability to it, through the genes. The term congenital infection, which is also used, is likewise confusing.

Transovarian Infection. The phases of parasitic animals which pass from one host to another by means of the egg must necessarily be small enough to enter the egg. It is not surprising, therefore, that they are single-celled Protozoa. Some of them cause serious diseases of animals useful to man. The first instance of this method of infection of the host was discovered by Pasteur in 1870, when he showed that this is the way in which silkworms are infected with the cause of the disease called *pébrine*. Silkworms are the caterpillars of the moth *Bombyx* and *pébrine* is caused by a minute single-celled protozoon called *Nosema bombycis*. The protozoon invades all the tissues of the mother moth and especially its ovaries, so that it enters the eggs and passes in these from one generation to the next.

Some species of the protozoan genus *Babesia* also pass into the eggs of the ticks which are their definitive hosts. During the sexual cycle of *B. bigemina*, which causes Texas fever of cattle, and during that of *B. canis*, the cause of malignant jaundice of dogs, the babesias enter the eggs of the female ticks and pass from these into the larvae which develop from these eggs. From the larvae they pass into the nymphs and from these into the adults. Thus the larvae, nymphs and adults of this generation of ticks derived from the infected eggs can all transmit these species of *Babesia*. The parent of this generation, however, cannot transmit them by its bite, because the babesias it sucks in all go into its eggs. From the adult of the generation thus infected, the babesias can pass to the larvae, nymphs and adults of the next generation also and in this manner some species of *Babesia* can persist in the larvae, nymphs and adults of as many as five generations of ticks. Their survival in the developmental phases of the ticks, like the survival of malarial parasites in mosquitoes (see Chapter 10), depends upon the temperature of, and probably also upon other factors in, the environments of the ticks.

Because the offspring of successive generations of ticks can thus become infective, the number of vertebrate intermediate hosts that may be infected is greater than it is when a parasite, such as the species of *Theileria* mentioned below, cannot thus infect the eggs of its definitive host. It will be greatest when the infected eggs are those of a three-host tick, such as *Dermacentor reticulatus*, the European vector of *B. canis*, because the larva, nymph and adult of three-host ticks each feed upon separate vertebrate hosts. It will be least when the infected eggs are those of a one-host tick, such as *Boophilus annulatus*, one of the vectors of *B. bigemina*, because the larva, nymph and adult of one-host ticks all feed upon the same vertebrate intermediate host.

Species of the genus *Anaplasma*, which is a relative of *Babesia* and causes the serious disease of cattle called anaplasmosis, can also pass through the eggs of the ticks which transmit them. Species of the related genus *Theileria*, on the other hand, which cause East Coast Fever and similar diseases of cattle, cannot do so; but, if the larvae of the ticks become infected with them, they can pass on to the nymphs and adult ticks.

Transplacental Infection. Infection of the young animal while it is still in the womb of the mother is seen most clearly among the mammals. It may happen either because the parasitic animal or its larva bores its way through the placenta by means of which the young animal obtains its nourishment from the walls of the womb, or because the parasitic animal or its larva enters the young in the mother's blood which circulates through the placenta into the young in the womb. It would be natural to expect, therefore, that all the animals which are parasitic in the blood of mammals, and also all those phases of other species which occur either in the blood or migrate about the maternal body, can pass in this way into the embryo while it is still in the womb. In fact, this is not so. The placenta seems to act like a protective barrier which excludes many of these species of parasitic animals, but it cannot prevent the entry of some others into the embryo in the womb.

Among the species which are parasitic in the blood and yet do not usually enter the child while it is still in the womb are the human malarial parasites. It is, in fact, known that there may be massive numbers of malarial parasites in the blood of the human mother without infection of the young in her womb. Yet some experts do believe that the unborn child may occasionally be infected with malaria by means of the mother's blood. Others believe that all instances in which this seems to happen are really instances of children being bitten by mosquitoes either while they are being born or immediately after birth. There do, however, seem to be some instances of infection of the child with malaria while it is still in the mother's womb. Probably this occurs only when the placenta has been damaged by disease or injury, so that its normal protection of the young in the womb has broken down.

Among species parasitic in the blood which may rarely pass through the placenta into the embryo is *Trypanosoma gambiense*, the cause of human African sleeping sickness.

Another organism which can pass from the mother into the young while they are still in the womb is *Bartonella bacilliformis*, which causes the South American (Peruvian) disease called verruga peruana and South American Oroya fever. Species of the genus *Toxoplasma* may also infect the young in this way, and there is one record of transmission of the human disease called kala-azar by this route, which occurred, curiously enough, not in India nor in any of the other countries in which this disease is endemic, but in England, where it does not normally occur.

Among the developmental phases of parasitic animals which are carried about the host's body in the blood, but are not actually parasitic in the blood, there are several which may enter the young in the womb by way of the placental blood stream. Among them are the later larval phases of the human blood fluke *Schistosoma japonicum*, those of the liver fluke of sheep and cattle, *Fasciola hepatica*, the bladder-worms of the tapeworms *Taenia multiceps* and *T. ovis*, and, among the roundworms, the larvae of *Habronema microstoma*, those of the lungworm, *Dictyocaulus filaria*, and perhaps also those of other species of lungworms, those of the dog hookworm, *Ancylostoma caninum*, and the human hookworm, *Necator americanus*, those of the filarioid heartworm of the dog, *Dirofilaria immitis*, and those of the ascarid of the dog, *Toxocara canis*, the large ascarid of man and pig, *Ascaris lumbricoides* and the ascarid of cattle, *A. vitulorum*. The larvae of *Ascaris lumbricoides* may migrate through the placental tissues into the embryos of pigs. The fact that its wandering larvae may also cause abscesses of the womb and vagina has already been mentioned. The larvae of this species may, indeed, turn up in almost any tissue of the body of the host.

An interesting arrest of the development of the infective larvae of the dog hookworm, *Ancylostoma caninum*, when they enter the embryo of the dog has been observed. It is an example of the correlation of the life histories of the parasitic animal and its host of which further examples are given in Chapter 6. Some 20 % of these larvae failed to develop further until the puppy under observation had actually been born. The same delay of the development of the larvae of *Ascaris vitulorum* may occur in the unborn calf.

CHAPTER 10

AVOIDANCE OF THE PARASITIC
ANIMAL BY THE HOST

Avoidance of Parasitic Animals by Hosts
other than Man

The word 'avoidance' in the title of this section needs a word of comment. Many biologists refuse, when they are discussing animals which are not human, to use terms like this one, which imply that there is a purpose in view; and certainly, if biological phenomena can be explained without the use of such terms, they should not be used. When, however, they exactly describe the phenomena in question and do not impute human characteristics to animals which cannot possess them, they are not objectionable, provided that their implications are understood. Certainly the term 'avoidance' seems to describe exactly the behaviour of some hosts to some kinds of parasitic animals.

Cattle, for instance, seem to know when the warble-fly is on the wing and that it will, if it can, lay its eggs upon their hairs. For the cattle then lash their tails and 'gad about', as the farmers say, in the manner described in Chapter 3, so that it is difficult to avoid the conclusion that the cattle are trying to keep these flies away. This kind of behaviour cannot be caused by fear of a bite or sting, because the mouth parts of the warble-flies are degenerate, so that they cannot bite. Nor do they possess any stinging apparatus. They have, indeed, no means of doing harm at all.

The direct cause of the marked change in the behaviour of the cattle may be the humming noise made by the warble-flies, but, if this is so, why do the cattle seek to avoid a fly which, whatever noise it makes, does not bite or sting? One of their other methods of avoiding these flies is by wading, if they can, into shallow water and by seeking whatever shade may be available. It seems as if they have learnt what men know, namely, that the warble-fly, like the tsetse fly which is one of its near relatives, avoids the shade and does not usually cross open water and that by immersing their feet and legs in water they protect the areas of their bodies on which the warble-flies most often lay their eggs.

Nor is this kind of behaviour confined to cattle which thus seem to try to avoid the warble-fly. Horses seem to try to avoid the horse bot-flies which lay their eggs upon them; and sheep, when the sheep nasal-fly is on the wing and is laying her larvae in the nostrils of these animals, try to avoid the fly by running about or huddling in groups with their nostrils pressed down between the bodies of their neighbours for protection. Both these kinds of fly belong to the same family as the warble-flies and, like the warble-flies, cannot bite or sting or otherwise injure the horses or sheep.

One feature of the kind of behaviour which has just been described is the *activity of the host*; and this may be the chief or the only means by which the host avoids the parasitic animal. Species of greenfly (Aphides), for instance, will walk away from species of insects belonging to the genus *Aphelinus* which are parasitic upon the greenfly; and the large maggots of some species of blowflies belonging to the genera *Lucilia* and *Calliphora* escape from the insect *Alysia manducator* by throwing it off and wriggling or burrowing into the decaying meat or similar material upon which these maggots feed. Both *Aphelina* and *Alysia* belong to the group of insects to which the ants, bees and wasps belong.

Other instances of the same kind of behaviour by insects could be given, and they certainly go far to convince us that the insects which behave in this manner are trying to avoid the parasitic animals concerned. This kind of behaviour is not to be confused with the quick or sudden movements of certain species of hosts which must present considerably difficult problems for some kinds of parasitic animals that must make contacts with them. Freshwater leeches, for instance, and the glochidial larvae of the swan mussel and other species which are parasitic upon freshwater fish, have to contend with the swift and often instantaneous changes of position of the fish.

Some kinds of hosts appear to resist certain kinds of parasitic animals in other ways. Thus the larvae of some moths belonging to the order Noctuidae often bite or shake off the maggots of the two-winged insect *Bonnetia comta*, which can become parasitic upon them. The larva of the codling moth has been seen in the act of biting and holding on to the female of a species of ichneumon fly belonging to the genus *Calliephialtes* and biting off part of the ovipositor with which this parasitic insect is trying to lay eggs upon the larva of the codling moth. The larva (pink boll-worm) of the moth *Gelechia (Pectinophora) gossypiella* may, before it is paralysed by the sting (ovipositor) of the ichneumonid species, *Microbracon mellitor*, bite off the organ with which this insect tries to lay its eggs upon the surface of the boll-worm.

This kind of behaviour has great biological interest. This interest is increased by the knowledge that, even after a parasitic insect has succeeded in laying its eggs upon its host, the natural movements of the host may dislodge these eggs or the parasitic larvae which develop from them. It is probable that all kinds of parasitic animals whose infective phases are attached to or penetrate through the outer covering of the host's body suffer the loss of many of these phases because they are removed by environmental agencies, such as currents of water round an aquatic host, by the bites and scratches of the host or by its use of fences, posts or other solid objects as rubbing posts. To meet this very risk, indeed, the infective phases of many parasitic animals which live on the surface of the host's skin have developed the hooklets, suckers and other means described in Chapter 5, or they are able to burrow into the skin so rapidly that the host has not time to prevent their entry into it.

Certain natural biological processes which the host must undergo may also interfere with the life of parasitic animals that attack it. Moulting (*ecdysis*) of the host's skin, for example, may do this. If, for example, the eggs of parasitic two-winged insects belonging to the family Tachinidae are laid upon the outside of the skin of caterpillars upon which these species are parasitic, these eggs are thrown off with the cast skin when the caterpillar moults. This is an instance of the kind of risk which compels the parasitic animal to adapt its life history to that of its host. Some of the ways in which various species of parasitic animals do this have been indicated in Chapter 6.

AVOIDANCE OF PARASITIC ANIMALS BY MAN

Man is able to avoid parasitic animals much more effectively than other animals can, because he can study, with his reason and intelligence, the modes of life of these parasitic animals and can therefore define precisely the ways in which they can make contacts with his tissues, so that he can try to prevent these contacts or at least to reduce them as much as possible. Man can, moreover, extend the protection afforded to his own species to other kinds of animals, such as his domesticated stock, which supply him with meat, milk, eggs and other useful commodities, such as hides and wool. A great part, indeed, of the work of the parasitologist is devoted to the study of means of preserving either man himself, or the supplies of food and other commodities which man derives from other animals, from the attacks made by parasitic animals.

When, therefore, we study man's avoidance of the parasitic animal, we enter upon a field which extends, for one reason or another, over pretty well the whole range of human activities. It leads us from the minor tragedies of the mosquito bite to the sufferings of human beings afflicted with hookworm infections in India, China, or the southern United States, to the prevention of war by the malarial parasite; from the monstrous injuries inflicted by the blood flukes upon many hundreds of human beings every year to the huge losses suffered by farmers all over the world as a result of the activities of the sheep stomachworm and its near relatives. A fascinating book could be written about the dramas, tragedies and tales of human and animal endurance which the study of man's battles with parasitic animals opens up. In this book, however, we cannot do more than indicate some of the principles upon which these battles are conducted. The reader will no doubt remember, as he reads the section which follows, further illustrations of these principles which have been given in earlier chapters.

The basic essential of any campaign against a parasitic animal is a thorough knowledge of every phase of its life history and also of its relationships with all the hosts in which it can live. We need to know all the hosts, because some of them may act as the reservoir hosts mentioned in Chapter 7, which may maintain sources of the parasitic animal from which man or his domesticated animals may become infected. When we have this knowledge of the parasitic animal and all its hosts, we can study the relationships involved and, guided by this study, we can select for attack the weakest points in the life history and biology of the parasitic animal.

The weakness or strength of the parasitic animal, or, in other words, its vulnerability, will vary with the degree of its dependence upon parasitic life. The facultative parasitic animal, for instance, which need not be parasitic at all, will be better able to resist attacks made upon it than the obligatory parasitic animal which must pass some part of its life history inside the body of some other animal. Among the obligatory parasitic animals there are species which need both definitive and intermediate hosts and these will, on the whole, be less able to resist the attacks of man, because man can subject them to a double attack directed not only against the parasitic animal, but also against its intermediate host, without which this kind of parasitic animal cannot live. If, for instance, we can destroy all the species of aquatic snails in which the blood flukes of man or the liver fluke of sheep and cattle must pass part of their life histories, we shall prevent the diseases which these flukes can cause.

When, however, species of this kind use more than one species of animal as their intermediate hosts, or more than one species as their definitive hosts, or more than one species of both definitive and intermediate hosts, our task will be correspondingly more difficult. The human blood fluke, prevalent in the Japanese area, uses man as only one of its definitive hosts; it can also use as definitive hosts the other animals mentioned below, and as intermediate hosts more than one species of aquatic snail.

If the parasitic animal does not use an intermediate host, the weakest point in its life history will usually be either its eggs or one or other of the larval phases which live a non-parasitic life outside the host before they reach the infective phase. Attacks can therefore be made upon these non-parasitic phases or they can be concentrated on the infective phase.

We can illustrate the principles of attack upon parasitic animals by considering briefly how they are applied in the fight against four species of parasitic animals which cause diseases of human beings, namely, the hookworm, the pork trichina-worm, the blood fluke and the malarial parasite. The two latter of these species use intermediate hosts, while the two former do not; and the early larval phases of one of each pair enter the world outside the host.

(1) *Avoidance of the Human Hookworm*

The life history of these species has been described in Chapter 3, in which it was explained that the larvae develop outside the host in damp soil until they reach the third stage, and that this third infective larva is the only larval stage that can infect man. It can enter the body of man either through the mouth or by penetrating his skin. Usually it enters through the skin. Two classes of factors can therefore prevent or limit the contacts between the host and the parasitic animal upon which the parasitic life of the hookworms depends. These are: (a) any influences which will prevent the development of the infective larvae; (b) any influences which will prevent the entry of these larvae through the human mouth or skin.

(a) Outstanding among the influences which can limit or prevent the development of the infective larvae are *climatic conditions*. The infective larvae develop from eggs passed out in human solid excreta, and the eggs and the first, second and third (infective) larvae are not parasitic. They are exposed to the climatic conditions which affect the earth in which they live. They can only survive and develop between certain limits of temperature and moisture. Periods of drought or spells of cold or of very hot and sunny weather may be sufficient to kill large numbers of the eggs and larvae. Even if these climatic conditions do not kill them, they may be powerful enough to render the infective larvae either inactive, or so feebly active that they cannot accomplish the penetration of the human skin. The larvae of the hookworms develop best in warm, damp soils that are free from salt or other chemical influences which are injurious to them. Vegetation growing in the soils may favour their development and survival because it shelters them from the injurious effects of heat, sunlight and drying.

The effects of climatic conditions are responsible for the restrictions of human hookworms to parts of the world in which the climatic conditions necessary for the development of the non-parasitic larvae prevail. These parts of the world are the Mediterranean region of Europe, South America, the southern United States, India, Egypt, central and south-east Africa, the Far East generally and north-eastern Australia and Melanesia and Polynesia. Hookworm disease may, however, occur in colder countries in certain situations, such as mines, in which the soil is damp and warm enough to enable hookworm larvae to develop to the infective phase. It used to be common in German mines and also in those of Cornwall, until hygienic measures eliminated it.

(*b*) The prevention of the entry of the infective larvae through the human mouth or skin is the most effective method of controlling the disease caused by these round-worms. It is all the more effective because, although these worms can be parasitic in a few animals other than man, man is their chief host, and practically all human infec-tions with them are acquired from other human beings. Man therefore holds in his own hands the abolition of this serious scourge, and he has, in fact, abolished hookworm disease from many areas simply by the hygienic measures which he has adopted.

He has prevented the pollution of the soil with human excreta containing the hookworm eggs, so that infective larvae cannot develop in it; he has prevented the con-tamination of his fingers and food and drink and eating and drinking utensils with the eggs and infective larvae; and he has prevented the contact of the infective larvae with his naked skin. If all human beings would adopt and maintain these relatively simple hygienic measures, hookworm disease of man would disappear from the world.

But although it is relatively easy to abolish it from civi-lised communities by the application of these hygienic measures, especially when the reason for them is explained to the people, it is not so easy to do this in parts of the world in which large numbers of people, most of whom are poor, ill-nourished, inadequately clad, badly educated and below the average level of human intelligence, live in poorly constructed houses, with inadequate sanitary arrangements, and work with bare feet and ankles in plantations, mines and other places in which the warm, moist soil is contaminated with human excreta containing hookworm eggs and provides a favourable medium for the development of the infective larvae. Under such conditions the human skin, especially the naked skin of the feet and ankles, is continually exposed to the infective larvae of the hookworms and the infection of many people is practically certain sooner or later.

(2) *Avoidance of the Pork Trichina-worm*

The avoidance by man of this cosmopolitan species of roundworm has been selected partly because it illustrates the exclusion of climatic factors that may help man in his war against the human hookworm. Climatic factors cannot operate directly against the pork trichina-worm, because this species never enters the world outside its many hosts. Its larvae find their way into the voluntary muscles of the host in which they are produced and there become enclosed in capsules (*cysts*), out of which they cannot make their way by their own efforts. They can leave these capsules only when the muscles in which they are imprisoned are eaten by a new host. The digestive juices of this host then digest the muscles and the capsules which imprison the larvae, so that they are released and can grow up in the intestine of the new host into their adult phases.

All hosts of the pork trichina-worm therefore infect themselves only by eating the muscles of other hosts that have also been infected. If man wishes to prevent all risk of his infection with the pork trichina-worm, all he has to do is to avoid eating living larvae in the flesh of any animal which may be infected with this species. How can he do this? He can either try to prevent the infection of any hosts of the pork trichina-worm which he eats as food, or he can cook, refrigerate or otherwise treat all meat derived from these hosts in such a way that any larvae of the pork trichina-worm which may be in them are all killed. The control of infection of man with the pork trichina-worm seems, therefore, to be straightforward and comparatively simple. In practice, however, it is not so simple as it seems to be.

The commonest source of human infection with the pork trichina-worm is the pig, and this is the reason why this species has been called the pork trichina-worm. Man usually infects himself by eating infected pork, for the simple reason that the pig is the host of this roundworm which is most often eaten by man. Man can infect himself just as readily by eating any other host of this species. One serious outbreak of the disease (*trichiniasis*)which the pork trichina-worm causes was, in fact, caused by the consumption in Germany of infected bear meat. The fact is, however, that practically all the infections of man which occur nowadays are derived from pork and especially from sausages of various kinds. The problem of the control of human infection therefore resolves itself for practical purposes into the prevention of the infection of pigs intended for human consumption and the adequate cooking, refrigeration or processing of all forms of pig-meat.

The prevention of infection of the pig is difficult, because the food of the pig is not easily controlled. Pigs, like man, can infect themselves only by eating flesh already infected, and a common source of infection of the pig is the rat. It is not easy, unless special and expensive measures are taken, to exclude all rats from quarters inhabited by pigs, and the entry into them of only one heavily infected rat may suffice to infect a number of pigs, and one infected pig, especially if its flesh is distributed among large numbers of sausages, can infect many human beings. Pigs may also be infected by the uncooked infected flesh of hosts other than the rat, so that it is good practice to feed to pigs only garbage or other food that has been well cooked. More effective still is the feeding of pigs on grain or other cereal or processed vegetable foods. But if the infection is to be controlled in every part of any particular country, these practices must be carried out carefully by every keeper of pigs, and it is practically impossible to ensure that this will be done. It is difficult, even in well-governed countries such as the United States, because the central government of the whole country is not able to control the regulations laid down by individual States, and both the central and the State governments may fail to control the actions of all the individual pig breeders. However careful the control of pigs may be, moreover, the habits of the pigs and other causes may fail to prevent their infection with the pork trichina-worm.

A second line of defence is therefore employed by many countries. This is the inspection of all pork intended for human consumption. The larvae of the pork trichina-worm (fig. 73) can, when they are magnified, be seen in the muscles of the pig, so that any carcasses containing them can be condemned. Usually small portions of one of the muscles which are most frequently infected—the midriff is the one usually selected—are squeezed between two pieces of glass in an apparatus called a trichinoscope, and the muscle thus flattened out is either examined under a low magnification of the microscope or its image is thrown on to an illuminated screen. This method takes up a good deal of time and skilled inspectors must be employed to carry it out. It is often, moreover, impossible to examine in this way the flesh of every pig being slaughtered, so that individual pigs which are infected may be missed. Further, if the infection is only a light one and the larvae present in the pig are few, none may be present in the small pieces of pork examined, so that, for this reason also, this method may fail. It does, however, detect many infected pigs and thus prevents the infection of many human beings.

Because this method of meat inspection is expensive and not by any means infallible, the authorities of some countries substitute for it, or supplement it with, one or other or both of two other methods. The first of these digests the samples of pork being examined in an artificial digestive juice and then examines the digest for the larvae of the pork trichina-worm. This method also consumes time and requires costly skilled labour.

The second method is practised by the authorities of the United States. It selects random samples of pork or pork products, whether these be canned or otherwise processed, and feeds them to rats. If there are any living larvae in the processed pork being examined, these will infect the rats and will grow into adult trichina-worms, from which larvae will be derived which will infect the muscles of the rats. When time has been allowed for this to happen the rats are killed, their carcasses are digested in artificial digestive juices and any larvae of the pork trichina-worm present in them can be recovered from the digests. This is the most expensive method of all, and it takes some weeks and the results of the examination are therefore delayed.

The third line of defence indicated above against human infection with the pork trichina-worm, namely, the killing of the larvae in the infected meat by adequate cooling or refrigeration, is the cheapest and most efficient one. It requires, however, care and attention to detail and, like the other methods mentioned, it may fail. The infected meat must be raised to a temperature of 58.3° C. (137° F.) if all the larvae are to be killed; or it must be cooled to a temperature of at least *minus* 15° C. and must be kept at this temperature for not less than 20 days.

When, however, cooking or refrigeration are being relied upon to kill all the larvae in the meat, it is essential to remember that every part of the meat must be heated or cooled to the temperature just stated. Ordinary kitchen roasting of a joint of pork, for instance, may raise the outer portions of the joint to a temperature which kills all the larvae, but the deeper portions of the joint may not reach this temperature, or this temperature may not be maintained in them for a sufficient length of time. A joint of pork roasted in the kitchen was, in fact, the source of one small outbreak of trichiniasis which occurred in England a few years ago. The regulations imposed by some countries nowadays state that only pieces of pork of the sizes specified may be refrigerated or otherwise processed. The importance of cooking pork was exemplified, during the recent World War, by the failure of a German commando unit operating on the Polish frontier region to cook it at all. These commandos halted for the night and commandeered a pig from a local farmer. They ate the pork raw and were, after a few days, all put out of action by trichiniasis. A similar example was provided by a German battalion stationed in Norway who were given pork for a meal. They ate it raw, and practically the whole battalion was put out of action by trichiniasis. Because many of the men sent gifts of the pork to their friends and relatives in Germany, these people also suffered from trichiniasis.

Joints of pork, or any other large pieces of it, are, however, less frequently the source of trichiniasis than are sausages. Not only may the flesh of one infected pig be distributed in sausages among a large number of people, but sausage-meat is eaten raw much more frequently than most people realise. The importance of cooking sausages thoroughly and of seeing that all parts of them are thoroughly cooked is very well shown by the outbreak of trichiniasis which occurred in Wolverhampton and Birmingham in the spring of 1941. This is one of the few major outbreaks of this disease that have occurred in this country, and the study of it revealed many interesting facts. Most of the people infected were factory girls who had not time to prepare a cooked meal to be taken with them to work. They took instead sandwiches quickly made by putting raw sausage-meat between slices of bread. Other workers in the same area who cooked their sausages were not infected; but many of the mothers contracted trichiniasis, either because they interrupted the cooking of sausages and licked their fingers before attending to some urgent task or other, or because they tasted the sausages to see whether they were sufficiently cooked. About half the people in the area who were asked whether they ever ate raw sausage-meat said that they did so; and other smaller outbreaks which occurred in England at about the same time revealed the fact that this habit of eating raw sausage-meat is not by any means confined to people in any particular walk of life.

The descriptions just given illustrate some of the methods

by which man avoids infection with parasitic animals which do not use intermediate hosts. The remaining examples to be given illustrate some of the methods that are used for the control of parasitic animals which use intermediate hosts.

(3) *Avoidance of the Blood Flukes of Man*

The life histories of the three species of blood flukes, all of which belong to the genus *Schistosoma*, have been mentioned elsewhere in this book. They follow, in general, the plan of the life history of the liver fluke of sheep and cattle described in Chapter 4. Their eggs are passed out in human excreta and the miracidial larvae which emerge from the eggs must enter certain species of aquatic snails, in whose bodies they must develop into larval phases similar to those of the liver fluke of sheep and cattle. These phases culminate in the cercaria (fig. 45) which leaves the snail and enters man, either through his mouth, or by penetrating through his skin.

These blood flukes, therefore, live only in those parts of the world in which their snail intermediate hosts also exist. Because these snails are aquatic and also require the warmth and other climatic conditions of the tropical and semi-tropical parts of the world, these blood flukes can only exist in these regions, and only in parts of these countries which provide the stagnant or slow-running water sheltered by vegetation in which the snails live. Actually the three species of blood flukes which infect man are limited, apart from one or two localised areas in Portugal, to northern and parts of eastern South America, to central and South Africa and its Mediterranean coast, to the course of the Tigris and Euphrates and some other parts of Arabia, and to Japan and parts of China and the adjacent region.

The influences which man can contribute to the prevention of his infection with blood flukes are not unlike those which he can practise for the prevention of his infection with hookworms. The cercariae of these blood flukes can, like the infective larvae of the hookworms, enter man either through his mouth or by penetration of his skin. Usually infection occurs by penetration of the skin. It follows that, if man (a) prevents infection of the snails in which the cercariae must develop, (b) destroys the snails and (c) prevents exposure of his naked skin or of his food canal to the cercariae which leave the snails, he will avoid infection with the blood flukes. How can he effect these aims? There are three main methods by which they may be attained.

(a) Man can avoid the contamination of all natural waters by his excreta. In this way he will prevent the entry of blood fluke eggs in his excreta from entering the surroundings of any snail intermediate hosts which may be present in these waters, so that these snails will not be infected with the miracidia of the blood flukes, and no cercariae will develop in the snails.

This method may accomplish a great deal. Most of the infections of man with the blood flukes *Schistosoma haematobium* and *S. mansoni* are derived from other human beings, because other animals are not usually definitive hosts of these two species. *S. japonicum*, however, is often parasitic, not only in man, but also in many other domesticated animals, including dogs and cats and in wild animals also, so that the excreta of these animals also may provide miracidia which infect snails and so contribute to the infection of man.

(b) Man can try to exterminate the particular species of snails which act as intermediate hosts of the blood flukes. This method also can accomplish a great deal. The snails are attacked by the cleansing of all the pools, waterways and other damp situations in which they live, by the removal of the vegetation from these waterways and pools under which the snails take shelter and by treating the waters with lime, copper sulphate, copper carbonate or some other chemical agent which will kill the snails. Treatment of the water is less effective against the snails which are the intermediate hosts of *S. japonicum*, because these snails are amphibious. In some small areas something can also be accomplished by the introduction into the waterways of ducks and geese which feed upon the snails and may appreciably reduce their numbers.

In countries such as Venezuela and Algeria, in which artificial waterways are used to irrigate the land, the irrigation canals provide excellent conditions for the breeding of snail intermediate hosts, so that they are prolific sources of human blood-fluke disease. Children play in and around these canals, their mothers use the water for washing purposes, and workers who operate the irrigation system must perforce go into the water or wet the naked skin of their hands and feet with it. Because the sanitary habits of the people are primitive, snails become infected by miracidia derived from human excreta which contaminate the waterways, and all the conditions necessary for the maintenance of the necessary contacts between the parasitic animal and its definitive and intermediate hosts are readily fulfilled. The destruction of the snails in the waterways, by periodic draining of the canals to expose the snails to drying, or the treatment of the water in them with chemicals, helps to break the chain of contacts; and, although this is difficult and expensive, much benefit results from it if the measures taken are energetic and are continuously maintained.

(c) In regions in which blood flukes and their intermediate hosts exist man can avoid the exposure of his food canal and of his naked skin to water derived from any kind of natural source.

Protection of his food canal is accomplished by drinking water only after it has been either boiled or treated with chlorine or some other agent which will kill the cercariae. Various kinds of filter have also been devised which will, it is claimed, filter out the cercariae.

But the wise man will not rely on the efficiency of water filters. He will boil all the drinking water and will take care also that unboiled water never enters his mouth or makes contact with his food. The skin can be protected by avoiding contact with natural or unboiled water. This involves the avoidance of bathing, shaving or other forms of immersion of the skin, wearing rubber boots and gloves, cleansing cooking and eating and drinking utensils with water which has been boiled, and similar measures. The life of the wise man in parts of the world in which these dangerous blood flukes exist will, in fact, be guided by a thorough knowledge of the ways in which they may make contacts with the snails which are their intermediate hosts and with man in whom they slowly cause, often over a period of many years, their dreadful and crippling effects.

(4) *Avoidance of the Malarial Parasites*

The malarial parasites offer us different and perhaps even more instructive examples of the factors which prevent or limit contacts between the parasitic animal and its intermediate and definitive hosts. In this instance man is the intermediate host, and his avoidance of the parasitic animal resolves itself largely into an avoidance of the blood-sucking insect which is the definitive host. A complete account of the various methods by which man seeks to avoid the malarial parasite would fill a lengthy book. Here we must be content to correlate man's methods of avoiding the malarial parasites with measures which he takes against the other species which have just been discussed. The basic principle which governs the fight against malaria is the same as that which controls our measures against the blood flukes. Apart from attacks upon the malarial parasites while they are in human blood, man tries in every way possible to avoid contact with the mosquito definitive host and also to destroy this host and to prevent it from breeding.

The available methods of avoiding the mosquito include the use of mosquito nets and of chemical substances which will, men hope, repel the mosquitoes, and also such planned methods as the construction of mosquito-proof dwellings and buildings, the placing of houses and other buildings away from the breeding-places of the mosquitoes, and the destruction of these breeding-places by drainage and other methods.

The mosquitoes themselves are attacked by means of chemical substances which will kill them, among which D.D.T. has become one of the most valuable. Their aquatic larvae, which must breathe air, are also attacked by covering the surface of the water in which they may be living with oils which exclude the air, or by treating the water itself with substances, such as Paris Green, which kill the larvae. The remarkable recent success of these attacks upon the mosquitoes and their larvae is well known and needs no description here. Their elimination from considerable areas of the world is one of the major accomplishments of modern civilisation.

It was in 1901 that Sir Ronald Ross, who proved that Sir Patrick Manson's theory that malaria is transmitted by mosquitoes is correct, recorded his opinion that it was silly to suppose that vast tracts of country could be freed from any mosquito; and no doubt most experts at that time would have agreed with him. But 30 years later, when fast French destroyers, and perhaps aeroplanes also, carried from Africa to Brazil the mosquito *Anopheles gambiae*, which is the chief transmitter of malaria in Africa, and thus made possible the occurrence in Brazil of one of the most severe outbreaks of malaria known to history (see Chapter 11), a team of American workers, financed by the Rockefeller Foundation and working with the Brazilian Government, was able to eliminate this mosquito entirely from the whole of Brazil. The methods used by this team of workers and the experience gained in Brazil have since been applied with equal success to the elimination of the same species of mosquito from Egypt, where, if it had not been eliminated, it would have taken a terrible toll of human life. Recently other areas have been freed from malaria-carrying mosquitoes and there is nowadays a real hope that malaria will eventually be brought entirely under human control.

At one time it was prevalent in England. It is said that both James I and Cromwell died from its effects, and many English people have suffered from it in the past. Even nowadays a few people who have never left England contract it. But it is not now an English disease. Why is it not? A brief discussion of the reasons why it is not will help us to realise that, although climatic factors undoubtedly play an important part in determining where malaria shall occur and where it shall not, these factors are not by any means the only ones which determine the incidence of this disease. But, before we try to answer the question why malaria no longer flourishes in England, let us enumerate briefly some of the most important conditions which must be fulfilled before the malarial parasite can establish itself in man, its intermediate host.

(i) The mosquito definitive host must take up from human blood mother cells (*gametocytes*) of the female sex elements (female *gametes*) of the malarial parasite. At certain times, however, gametocytes may be absent from the blood of people infected with the malarial parasites.

(ii) The gametocytes must be a certain number of days old, because, if they are not, they may fail to develop in the mosquito.

(iii) Enough of the gametocytes must be taken up to establish an infection in the mosquito.

(iv) A sufficient number of male, as well as female, gametocytes must be taken up, so that enough female gametes may be fertilised by male gametes.

(v) The temperature of the environment of the mosquito must remain within limits between which the development in it of the malarial parasite can occur; and it must remain between these limits for the period of time required for the completion of the sexual development (*sporogony*) of the parasite. The sporogony of *Plasmodium vivax* is completed in 16 days when the external temperature is 20° C. and in 17–18 days when it is 23–28° C. The sporogony of *P. malariae* is completed in 30–35 days when the external temperature is 20° C. and in 28 days when the external temperature is 23–28° C. The sporogony of *P. falciparum* is completed in 10–11 days when the external temperature is 30° C., in 17–18 when it is 23–28° C. and in 22 days when it is 20° C. *P. vivax* can develop in mosquitoes at lower temperatures than the other species can and this is one of the reasons why it is able to extend into temperate regions, while *P. falciparum* develops best in warmer countries. For the same reasons malarial infections are not usually acquired during the winter in temperate countries, but may be acquired all the year round in the tropics. If the temperature falls below 16° C., the sporogony of all the species of malarial parasites is usually not completed and zygotes are not formed.

Some varieties of malaria-carrying mosquitoes, moreover, breed in brackish water, while others breed in fresh water. This is one reason why *Anopheles maculipennis* var. *atroparvus* is the only important vector of malaria in northern Holland. It breeds in the brackish waters there. In southern Holland, on the other hand, the waters are relatively fresh, and the mosquitoes which breed in them are not vectors of malaria.

(vi) The climatic conditions must be suitable for the completion of the life history of the mosquito. Mosquitoes must lay eggs in water in which their larvae must develop. Malaria is therefore commoner at low altitudes, where collections of stagnant water favour the development of the mosquitoes. The altitude by itself probably has little effect upon the malarial parasites. Human occupations, such as excavations, building, bad farming and so on, which encourage the creation of suitable breeding pools help the development of mosquitoes.

Another important detail is the fact that mosquitoes bite chiefly at dusk or at night and during the day shelter in dark places. In daytime, therefore, they may lurk in buildings, or under vegetation, the growth of which is greater in warm and humid climates.

(vii) If the climatic conditions are not suitable for the mosquitoes, they may die before the sporogony of the malarial parasites can be completed in them. They require, for example, a certain degree of humidity of the atmosphere; if this is too low, many of them will die. It was found, for example, that in England in April and May less than 10 % of *A. maculipennis*, which is the chief vector of malaria in Britain, lived long enough to produce zygotes, so that only this proportion of them became able to infect man. In August and September, on the other hand, 50% of them could live long enough to become infective.

(viii) Climatic conditions may profoundly affect the density of the mosquito population. Unless there is a sufficient number of malaria-carrying mosquitoes, serious outbreaks of malaria will not occur. During the winter in the southern United States the climatic conditions reduce the number of living mosquitoes.

(ix) Malaria cannot be transmitted to man unless the mosquitoes which can transmit it make contact with man and suck his blood. The habits of both man and the mosquito therefore play an important part in the transmission of malaria and both are affected by climatic conditions. Man exposes his skin more frequently and more extensively in warm or hot weather, when the mosquitoes are more likely to be infective; during cold spells in summer he may retire into houses and buildings in which infective mosquitoes are also taking shelter. Climatic factors may make possible certain human occupations that involve life in camps, or work in fields or plantations, which may increase the opportunities of the mosquito to bite man.

The habits of the mosquito are even more important. Within some species of mosquitoes biological races have been distinguished, whose feeding habits are different. Within the European species *A. maculipennis*, for example, which is the chief vector of malaria in Britain and the northern provinces of Holland and the most important vector in Europe and North Africa, there are several races, of which the race *atroparvus* mentioned above is an example, which bite man only exceptionally; they prefer the blood of animals, rather than that of man, as, indeed, this species as a whole normally does. This partly explains why it may happen that malaria does not occur in areas in which *A. maculipennis* is abundant, while in other nearby areas more malaria occurs, although this species of mosquito is less numerous. In the former areas the more abundant mosquitoes belong to races which do not bite man, while in the latter areas they belong to races which do Another factor also comes into play. The females of *A. maculipennis*, although they prefer animal rather than human blood and are more numerous in the stables and sheds of domesticated animals, readily enter human dwellings. This habit gives them better opportunities of sucking human blood and of communicating malaria to man, who is usually bitten at night. In the autumn these mosquitoes may be more numerous in human dwellings and they may then, provided that the temperature allows the malarial parasites to survive in them, transmit malaria to man during the winter.

(x) Finally, it is known that some of the biological races which are distinguished within species of mosquitoes are more readily infected with some species of malarial parasites, or with some strains of them, than other races are. Thus English workers found that *A. maculipennis* in England could be infected with a strain of *Plasmodium falciparum* obtained from human beings from Italy, but not with a strain obtained from human beings in India. Some races of mosquitoes can, moreover, transmit some species or strains of malarial parasites by a single bite, while several bites are required to establish other strains or species of malaria in man.

The factors which have just been outlined largely explain the geographical distribution of malaria and, in particular, its occurrence in England. British-born mosquitoes belonging to the species *Anopheles maculipennis* cannot infect people living in England unless they infect themselves by sucking human blood containing malarial parasites. They may, for example, suck the blood of people who have just returned from tropical countries in which they have acquired malarial infections. The climatic factors which have just been described then come into action. Fortunately for English people, the temperatures necessary for the completion of the sporogony of the malarial parasites occur only during the summer months and then do not frequently persist for long enough to enable the sporogony to be completed. This is one of the reasons why malaria has not become firmly established in Britain. Another reason is the fact, already mentioned, that malaria does not become regularly established (endemic) in any area unless there is, in that area, a sufficient number (population density) of the right species of mosquito.

From time to time, however, there are, as we have already noted, a few reports of truly indigenous malaria in Britain, of people, that is to say, who have contracted malaria which has been transmitted from one person living in Britain to another and not to British people from persons who have recently returned from malarial districts abroad.

More usually when malaria occurs in British people who have not had it before, it has been conveyed to these people by British-born mosquitoes which have recently bitten a person who has contracted malaria abroad and when local British climatic conditions have allowed the malarial parasites to complete that part of their life history which occurs in the mosquito. This has happened especially when relatively large numbers of persons infected abroad have come into Britain. Often these persons have been soldiers returning home after service overseas.

After the 1914–18 war, for instance, outbreaks of malaria occurred in Britain which were caused in this way. Similar outbreaks then occurred in Germany, Holland and some other European countries. In northern Holland the infection was maintained by healthy carriers of it, especially children. Most of the British people infected were those who lived in the low-lying areas of East Anglia and along the south-east and south coasts of Britain as far as the Isle of Wight, where *A. maculipennis* is most often found, although there were also isolated areas in Gloucestershire on the Severn Channel, and in the Morecambe Bay district, where the necessary climatic conditions also allowed smaller outbreaks to occur. In the sixteenth century the distribution of malaria in England was much the same, although at that time there were malarial areas also in the East Riding of Yorkshire and in the Solway district.

The species of malarial parasite most often acquired in Britain is *Plasmodium vivax*, because this species can complete that part of its life history which occurs in the mosquito at atmospheric temperatures lower than those which are required by the species which causes malignant tertian malaria. This latter species, *P. falciparum*, readily infects British *Anopheles maculipennis*, but usually it dies before its sporogony in this mosquito is completed. *P. vivax* may remain inactive in man for some months. If, therefore, a man is infected with this species by a mosquito which sucks his blood in the autumn, the parasite may not become active enough to cause symptoms of malaria until the following spring or summer. By that time the parasites may have died out of the mosquitoes, so that it may be very difficult to understand how these attacks of malaria which occur in the spring or summer have been produced. It is, however, important to remember that the parasites may survive in the mosquito for a certain time at external temperatures which are too low to allow their development to proceed, and that when the temperature rises sufficiently high, their development may proceed.

The foregoing account of man's avoidance of the malarial parasites is incomplete, but it will no doubt suffice to illustrate the main principles which guide man's battle with this particular parasitic animal. The reader will have noticed, no doubt, that nothing has been said about another feature of this battle which is important, namely, the attacks which are made upon the malarial parasites while they are in the blood of man. This method of attack upon the parasitic animal has been omitted also from the accounts of the other species which have been considered above. It has been omitted because it would involve the description of complex chemical substances, such as mepacrine, paludrine and other antimalarial compounds, the compounds of antimony which are used against the blood flukes and the hydrocarbons and other substances with which hookworm infections are treated. Descriptions of these compounds and their modes of action would take us far beyond the scope of this book. This method of direct chemical attack upon the parasitic animal while it is in the host is, moreover, less a method of avoiding it and more a method of combating it after avoidance of it has failed. But it is, of course, a very important method of attack. For it not only frees individual hosts from the parasitic animals concerned, or at any rate reduces the number of these to such a degree that the host can cope with them unaided, but it also reduces the degree to which the host can act as a source of the infections of other individual hosts. In this latter sense it can, indeed, be regarded as one of the methods of avoiding the parasitic animal and, when the virulence of the species concerned is not very high, this reduction of the number of sources of infection may be the only degree of control that is required.

CHAPTER 11

CONCLUSION

An attempt has been made in the preceding pages to bring to the notice of the reader the most important features of the relationships which exist between parasitic animals and their hosts. It has been necessary to refer only briefly to some aspects of these relationships and to omit references to others, but the selection and balance of the material available has been guided by several main objectives.

An attempt has been made, for instance, to show that host and parasitic animal must always be considered together, because the parasitic animal is, like all other living things, including man, intimately related throughout its existence to its environment. The fact that its environment is, for a part or the whole of its life, the surface or interior of another animal, does not absolve the parasitologist from the biologist's practice of considering animal and environment as a biological whole.

A second objective has been the demonstration that some species of parasitic animals are among the most powerful enemies of man and his civilisation. For this reason, many of the instances of parasitic life and its effects have been selected from the work of parasitologists who have studied the species of parasitic animals which injure man and his domesticated animals. Space has, however, been found for references to some species which do not affect man or his domesticated animals at all. These have been included because the parasitologist who studies species which are economically important must consider also the species which are not and must base his work upon the purely biological study of the host-parasite relationship. He is not one of those people who say: 'This animal is injurious to man; how can I exterminate it?' He studies the parasitic animal rather as a species which exhibits a particular way of life and, shedding his human spectacles so far as he can, tries to see it as a species whose biological status is equal to that of man himself.

It is not easy to do this. Human beings are swayed by emotions which do not affect the beautiful trypanosome when it kills our best-beloved, nor the hookworm when it sucks our children's blood and creates in them an anaemia which wilts their natural mental and physical development. Nor does that beautiful creature, the mosquito, consider its ways when it fills the blood of a Cromwell or James I with malarial parasites. It would have filled the blood of a Hitler no less effectively with them if geographical and other circumstances had given it the opportunity. It might have thus killed Keats, as tuberculosis did. It is possible that it did contribute to the death of Byron. The parasitic animal is not concerned with human values. It follows, as other animals must, the natural urges which impel all living things to get their food, mate with their kind and attempt to ensure the survival of their species. Some of the methods by which it does these things have been considered in this book. We shall miss a great deal of the beauty which these processes display, and most of the other beauties with which parasitic animals can delight our understanding, if we allow our studies of them to be influenced by disdain of the parasitic way of life, instinctive and powerful though that disdain may be.

For the beauties exhibited by parasitic animals must fill

us with a reverence no less than that which we feel when we contemplate the beauties of animals whose ways of life we can more readily approve. They are not confined to the marvels of structure which the eye perceives. More exhilarating than these are the beauties of physiological process and adaptation, the full delicacy and strength of which can hardly be appreciated until the mind has submitted, for a number of years, to those disciplines of study, emotional reaction and philosophical reflection which are rewarded, in the end, by a measure of understanding. These disciplines bring, if the mind can sustain them, rare moments of insight, moments during which the mind is reverently aware, through all its doubts and distresses, of a certainty of the unity of all phenomena appreciable by the human mind and of an inner reality which inspires and pervades the whole universe. It is this insight which equates the amoeba with the life and character of him who has thought his apprehension like a god's; it is this insight which humbles our pride and helps us to endure the vast sufferings which are inflicted by the parasitic animal. It has not been one of the primary objects of this book to dwell upon these sufferings. They have been implied, rather than specifically stated, in the preceding pages. It will not, however, be out of place at this point to indicate that the powers of the parasitic animal to disturb any notion that we may have that this universe has been created for the convenience of man are not confined to their inroads upon our philosophical complacencies. They are matters of hard and astonishing fact. The number of human beings infected by parasitic animals is, for instance, almost incredible. The American helminthologist, Dr Norman Stoll, has published startling estimates of the number of people infected with various kinds of parasitic worms in different parts of the world. He has calculated that 644 million people are infected with the large roundworm, *Ascaris lumbricoides*, and some 209 million people with the seat-worm *Enterobius vermicularis*. The incidence of infection with both these species is increasing in Europe under the present social conditions there. Dr Stoll estimates that 27 million people are infected with the pork trichina-worm, 456 millions with human hookworms, 114 millions with the blood flukes, 189 millions with Bancroft's filarial worm, 20 millions with its relative *Onchocerca volvulus*, and 13 millions with the related 'eye-worm', *Loa loa*. He calculates that some 39 millions are infected with the beef tapeworm, *Taenia saginata*, some 20 millions with the dwarf tapeworm, *Hymenolepis nana*, about 10 millions with the fish tapeworm, *Diphyllobothrium latum*, but only about 2½ millions with the pork tapeworm, *Taenia solium*. To these parasitic worms we must add the parasitic Protozoa, which include the amoeba which causes amoebic dysentery, *Entamoeba histolytica*, which is present in large numbers of people, although it causes disease in only about 10 % of them, and the trypanosomes and the malarial parasites, the incidence of which can be calculated only with difficulty, if at all. Malaria is certainly still one of the most serious epidemic diseases of man. In 1950 the World Health Organisation reported that every year it kills 3 million people and that each year 300 million new malarial infections occur. Malaria can cripple tropical industry and it has been responsible for many massacres of human beings. It is among the few agencies that can put an end to human wars. Malaria immobilised, for instance, the opposing armies in Macedonia in 1916. During the recent war in the Far East, the ratio of battle casualties to casualties caused by malaria, dengue fever, dysentery, scrub typhus and skin diseases was, between September 1943 and February 1944, 1 to 15. Of a total of 47,534 casualties caused by these diseases, 28,909 were caused by malaria.

Turning from man himself to his domesticated animals, upon which he depends for milk, eggs, meat and other products, we find that the losses of domesticated animals caused by parasitic animals are very great. The United States Bureau of Animal Industry estimated in 1942 that the various animals parasitic in farm stock caused, in the United States, a loss of 290 million dollars every year, a figure which is 69% of the total cost of all kinds of disease of farm stock in that country. In other countries the losses are, in proportion to the numbers of farm animals maintained, equally high, although in some parts of the world diseases caused by certain bacteria and viruses may account for a higher proportion of the harm inflicted. Certainly in Britain parasitic animals inflict heavy losses upon the farmer. To the losses of farm animals caused by them we must add the extensive injuries some of them do to vegetable foods. All the insect pests of crops are, in a sense, parasitic upon these crops and it is impossible to estimate the damage done to crops of wheat, potato, sugar-beet and some horticultural plants caused by the eelworms, which are roundworms.

There are, therefore, sound practical reasons for the expenditure, by all countries of the world, of large sums of money every year upon attempts to limit or remove the harm done by parasitic animals. Not for nothing do army commanders plan to defeat them and governments mobilise science against them. Our knowledge of them, gained for us by the labours of medical men, biologists and veterinarians, some of whom have been killed by the parasites they were studying, has given us more effective control of their activities. It would have been interesting to have written more about this never-ending war upon the parasitic animal, but the reader may find it interesting to work out for himself effective plans of campaign against the species of parasitic animals which have been described, remembering, as guiding principles, that the wise offensive strikes at the weakest point in the life history of the enemy animal and adapts the measures taken to the various environments of the hosts and parasitic animals concerned.

Two features of our modern civilisation which help the dissemination of some kinds of parasitic animals are especially important. These are the internal combustion engine and the aeroplane. The invention of the aeroplane in particular has provided some species of parasitic animals with effective, artificially warmed and very fast transport which can carry them rapidly about the world. Man has had more than one bitter experience of the dreadful efficiency with which this product of his ingenuity may bring death and suffering to thousands of his fellow men and women. One instance has already been briefly mentioned, namely, the transport in 1930 from Africa to South America, by aeroplanes and fast naval destroyers, of living specimens of the malaria-carrying mosquito, *Anopheles gambiae*, with the result that the city of Natal, in Brazil, experienced in 1930 and 1931 two severe outbreaks of malaria. Later, when this species of mosquito had been carried by ships, motor-cars and other agencies to an area of north-eastern Brazil which was thickly populated by poorly nourished and therefore feebly resistant people, there followed, in 1938, one of the worst outbreaks of malaria known to history. In June and July of that year some 20,000 Brazilians died of this disease.

If the fact that these battles are continuously necessary should make us unhappy, if we rebel against the necessity of recognising that the price that we pay for existence is a constant vigilance against other living things, we may reflect that this price must also be paid by the animals that we fight and by all other living things and that human life, which some consider to be the crowning product of the process that we call evolution, is not the final objective of that process. The parasitic animal may seem to us to be an enemy and an evil form of life, but we cannot deny that it is also an actor in the drama of the universe; and, if we admit so much, shall we deem it of lesser value than man himself? Divine values are not determined by man. It is, on the contrary, man's privilege to discern them in everything that he perceives. Whether we believe, with some philosophers, that these values are immanent in the universe that we know, or with other philosophers, that they exist outside the universe and are but dimly revealed to some of us to whom pain, or happiness, or a natural endowment, brings an insight into them, it is easy to discern them in the beautiful, the good and the true, more difficult to see them in error, ugliness and evil. Yet in these also they must reside, if the human passion for consistency is to be satisfied.

Perhaps this passion is itself an error; perhaps it is no more than the consequence of too exclusive an allegiance to logic, to scientific method, to impatience with disorder. All these impel the human mind to impose upon the ever-changing universe those categories which are the framework of its thought. The imperfections of this method are evident in every field of human activity. They are evident when the artist attempts to depict in static paint the inexhaustible colour and movement of the clouded, wind-swept skies; they are evident when the poet, better equipped with words, weaves in a tissue of meaning and music the ecstasies and distresses of his soul; they are evident when the musician, reaching after the harmonies of the spheres, brings echoes of them to those who have ears to hear. These imperfections are, some think, even more acutely perceptible in the truths which the scientist wrings by patient effort from the materials he studies. If the bird on the bough sing clearly of Heaven, the predatory carnivore, stealing upon it to strike it down, exhibits no less than its victim the lineaments of God. The parasitic animal, following its ways in the body of either of these creatures, must also reveal these lineaments as does also the man who observes, with anger and despair, its destruction of the beauty that he adores.

The dilemma, then, is this. We try, in our human pride and self-centredness, to find a God who shall have made a universe suited to our ideas. If beauty is our quest, we want it good and true; if truth is our aim, we find that it is always beautiful, although science is always changing it; and when we seek the good, the discovery of evil fills our souls with pain. The way out is the way of the great objective artist, the method of taking off our human spectacles, of conquering the self and entering into the souls of the objects which we perceive. Only then, when we know them from the inside, can we take on again our personal habiliments and recreate for other men the experience that we have gained. In this way Macbeth and Beatrice were born. In this way can be born, with all the integrity of human art, the parasite which will destroy, without consciousness of any philosophy, so much that man reveres as the handiwork of Heaven.

The attempt to do this, to enter into the non-human, whether it be living or not, and to recreate it, when it is understood, for the contemplation of our fellow-men, is the task of the saint, the artist and the philosopher rather than that of the biologist. He is, however, a poor biologist who does not try to be something of a seeker after God as well. To him the parasitic animal will bring an especially difficult challenge. With the meeting of this challenge this book is not concerned. Perhaps, if circumstances permit, it will be taken up in another book, a book more contemplative than this and less concerned with the practical urgencies of human and animal life and suffering.

INDEX

Printed in the United States
By Bookmasters